AN
INTRODUCTION TO
USING GIS IN MARINE BIOLOGY
SUPPLEMENTARY WORKBOOK ONE
Creating Maps Of Species Distribution

About The Author: Dr. Colin D. MacLeod graduated from the University of Glasgow with an honours degree in Zoology in 1994. He then spent a number of years outside of the official academic environment, working as, amongst other things, a professional juggler and magician to fund a research project conducting the first ever study of habitat preferences in a member of the genus *Mesoplodon*, a group of whales about which almost nothing was known at the time. He obtained a masters degree in marine and fisheries science from the University of Aberdeen in 1998 and completed a Ph.D. on the ecology of North Atlantic beaked whales in 2005, using techniques ranging from habitat modelling to stable isotope analysis. Since then he has spent time working as either a teaching or research fellow at the University of Aberdeen and has taught Geographic Information Systems (GIS) at the University of Aberdeen, the University of Bangor (as a guest lecturer) and elsewhere. He has been at the forefront of the use of habitat and species distribution modelling as a tool for studying and conserving cetaceans and other marine organisms and has co-authored over forty scientific papers on subjects as diverse as beaked whales, skuas, bats, lynx, climate change and testes mass allometry, many of which required the use of GIS. In 2011, he created *Pictish Beast Publications* to publish a series of books, such as this one, introducing life scientists to key practical skills and *GIS In Ecology*, to provide training and advice on the use of GIS in marine biology and ecology.

PSLS

AN INTRODUCTION TO
USING GIS IN MARINE BIOLOGY
SUPPLEMENTARY WORKBOOK ONE
Creating Maps Of Species Distribution

Colin D. MacLeod

2nd Edition

Pictish Beast
Publications

ISBN – 978-0-9568974-3-5.

Published by Pictish Beast Publications, Glasgow, UK.

Printed in the United Kingdom

First Printed: 2011; Second Edition: 2013

Trademarks

All terms mentioned in this book that are known to be trademarks or service marks have been appropriately capitalised. The author and the publisher cannot attest to the accuracy of this information. The use of a term in this book should not be regarded as affecting the validity of any trademark or service mark. In addition, the use of a trademark should not be taken to indicate that the owner of that trademark endorses the contents of this book in any way, or that the author and publisher of this book endorses a particular brand or product.

Warning And Disclaimer

Every effort has been made to make this book as complete and as accurate as possible, but no warranty or fitness is implied. The information provided is on an 'as is' basis and is provided as examples for training purposes only. The author and the publisher shall have neither liability nor responsibility to any persons or entity with respect to any loss or damages arising from the information contained in this book.

'Space, the final frontier...'
James T. Kirk,
Captain, USS *Enterprise* NCC 1701

This book is dedicated to those who wish to explore
all aspects of how space and spatial relationships influence
marine organisms and ecosystems, but don't know where to start.

Table of Contents

Preface

When I first started writing *An Introduction To Using GIS In Marine Biology*, one of the main aims was to provide marine biologists with both an introduction to using GIS in a language which they could understand and a reference guide of how to do specific tasks using GIS. However, it quickly became clear that while people would find this useful, they also wanted to have exercises which they could work through to develop their GIS skills. This was something which I had originally planned to include in *An Introduction To Using GIS In Marine Biology* but found I had run out of space if I wanted to be sure that the finished book could be lifted by a single person unaided. I also found that by including them within the main GIS book I could only cover a small number of exercises. Thus, I decided to create a number of separate supplementary workbooks, each of which would provide exercises on related topics which could be worked through using specific data sets. These exercises could then be updated as and when required, and new workbooks could be added to cover different sets of related tasks common to different areas of marine biology without the need to update the main, and more expensive, volume.

This is the first supplementary workbook and provides five exercises related to creating simple maps of species distribution and represents tasks that many marine biologists are likely to encounter in their daily research. Therefore, they represent basic GIS skills which marine biologists should be able to do with relative ease. The instructions for these exercises follow the structure for combining instruction sets for individual tasks provided in the '*How To...*' reference guide section of *An Introduction To Using GIS In Marine Biology* to complete more complex tasks, as outlined in chapter twenty of that book. By working through these tasks using the data sets provided, you will gain both experience in doing these tasks, and an insight into how you can create such combined instruction sets. In addition, while the instructions provided for each exercise are based on specific data sets, they can easily be adapted for use with other data sets as required.

--- *Chapter One* ---

Introduction

The aim of this workbook is to help marine biologists familiarise themselves with using GIS in their research. To do this, it uses a Task Oriented Learning (TOL) approach, first introduced in *An Introduction To Using GIS In Marine Biology*, to provide five exercises based around creating simple maps of species distribution, something which many marine biologists need to be able to do on a regular basis. As such, it does not represent a stand alone GIS book and is meant to act as a companion guide to the original book rather than to replace it in any way. It does not provide any background information on using GIS as this has already been covered within *An Introduction To Using GIS In Marine Biology* itself. Instead, it simply provides instructions on how to do the exercises themselves.

Thus, this workbook is primarily aimed at those who have read some or all of *An Introduction To Using GIS In Marine Biology*. If you have not already done so, it is recommended that at a minimum you read chapters seven (*Translating biological tasks into the language of GIS*), eleven (*How to use the 'How To...' sections of this book*) and twenty (*How to combine instruction sets for basic tasks to create instruction sets for more complex tasks*) of *An Introduction To Using GIS In Marine Biology* before working through any of these exercises. It will also help if you are familiar with the basics of GIS (chapter two), common concepts and terms in GIS (chapter three), the importance of projections, coordinate systems and datums (chapter four), types of GIS data layers (chapter five), starting a GIS project (chapter six) and how to set up a GIS project (chapter thirteen). Finally, it is worth at least flicking through chapters thirteen to nineteen to familiarise yourself with how instruction sets are laid out using the TOL approach introduced in *An Introduction To Using GIS In Marine Biology*.

This second edition of *Supplementary Workbook One* uses ArcGIS® 10.1 software to illustrate how these tasks should be done. However, similar processes are likely to be used to achieve similar outcomes in other GIS software.

The exercises provided in this book are designed to be worked through in a sequential manner. This is because the same data sets are used throughout and you will need to use some of the data layers generated in earlier exercises for later ones. In addition, the exercises lead on from each other in a manner that develops your familiarity with using GIS to map species distributions, so increasing the benefit of conducting the exercises in the sequence provided. For example, while the first exercise simply allows you to produce a map of where a species was recorded, later exercises show you how to produce more complex species distribution maps of species richness and ones which take account of the amount of survey effort in different parts of a study area.

The exercises are provided using the same flow diagram based format introduced in the *'How To...'* reference guide section of *An Introduction To Using GIS In Marine Biology*, and specifically in chapter twenty which outlined how to combine individual instruction sets to work out how to do more complex tasks. This means that for each exercise, you will first find an outline of what will be achieved by the end of it, why it is useful for marine biologists to be able to do this and what data layers you will need to start with. You will then find a summary flow diagram which will detail the order which individual instruction sets for basic tasks must be done. Finally, you will find a set of numbered instruction sets based on those provided in *An Introduction To Using GIS In Marine Biology*. These have been customised to make them specific to the data set used for each example. In order to complete a specific exercise, you will need to work through each of these instruction sets in the order given in the summary flow diagram. In order to allow you to know whether you are progressing correctly, figures will be provided at regular intervals which will show you what the contents of the MAP window, LAYOUT window, TABLE OF CONTENTS window and/or TABLE window should look like at that specific stage.

The data sets used in each exercise can be downloaded from www.gisinecology.com/books/marinebiologysupplementaryworkbook.

NOTE: The instruction sets provided here are for training purposes only, and are only meant to be an aid to learning how to use GIS in marine biological research. While every effort has been made to ensure that these instructions are complete and error-free, they come with no guarantee of accuracy and, as with all technical books, some errors may have slipped through undetected. Whenever I become aware of any such issues, I will post corrections on www.GISinEcology.com/books/marinebiology/corrections rather than waiting to correct them in the next edition of this book. As a result, before you do any of the exercises in this book, you should check this page to see whether there are any corrections that should be applied. In addition, it is important to realise that there is no guarantee that these instructions will produce the desired outcome in every circumstance. As a result, if you are using the instruction sets provided here to learn how to do critical tasks, it is essential that you check (and then double check) that they work for your given circumstances rather than blindly following them without thinking. The author will not be responsible for any errors which occur because of the application of these instruction sets to real world situations.

NOTE: As with many things in GIS, there may be more than one way to do the exercises outlined in this book. The instructions presented here will work for the data set provided and for the exercises outlined in this book. They should also work in most other circumstances. However, if you find an alternative way to do them which works for your data, or if you have someone who can show you how to do them in another way, feel free to do them differently.

Chapter Two

How To Use The ArcGIS 10.1 Software User Interface

In this book, ArcGIS 10.1 software is used to illustrate how the steps outlined in each instruction set can be applied to a specific GIS software package. All of the instructions provided in this section of the book assume that you are working with the main ArcGIS 10.1 user interface, a module known as ArcMap. As a result, it is important to outline how the ArcMap window is laid out.

The ArcMap user interface is divided into a number of sections and windows (figure 1). These sections include: a MAIN MENU BAR that allows you to access some basic and commonly used tools through a variety of standard drop down menus; the OPTIONAL TOOLBARS AREA, where you can display toolbars for specific toolboxes to allow you to access them quickly and easily; the TABLE OF CONTENTS window which displays a list of all the data layers in your GIS project; the TOOLBOX window, which allows you to access the various tools available in the software; an ADDITIONAL OPTIONAL TOOLBARS AREA; an X-Y COORDINATE DISPLAY area, which provides the X and Y coordinates for the position of the cursor in the MAP window; and the MAP window which displays all the active data layers within an active data frame.

When you open the ArcMap module for the first time, you may find that some of the windows (such as the TOOLBOX window) are not visible at all. This is because it can be customised depending on the user preferences. However, for the purposes of this book, it is useful to standardise this layout in terms of what is visible at all times and where the different windows are placed. To do this, first check if the TABLE OF CONTENTS window is visible. If it is not, go to the WINDOWS menu and select TABLE OF CONTENTS. If it is visible (or once you have made it visible), click on its title bar and hold the mouse button down. Now drag the TABLE OF CONTENTS window towards the middle of the MAP window. Four blue arrows will now be visible there. Drop the window onto the left hand arrow of this group. This will tie the TABLE OF CONTENTS window to the left-hand side of the MAP window. Next, click on the GEOPROCESSING menu and select ARCTOOLBOX. The TOOLBOX window will now appear at the left hand side of the TABLE OF CONTENTS window. Move it to the right of this window, by clicking on its title bar and holding the mouse button down. Now drag it to the left hand arrow of the central group and drop it there. The ArcMap 10.1 user interface will look like that shown in figure 1. It is advisable that you set up the ArcMap 10 user interface to look like this before doing the exercises in this book.

NOTE: It is useful at this time to also change two other default settings of the ArcMap module. Click on the GEOPROCESSING menu on the main menu bar and select GEOPROCESSING OPTIONS. Make sure that there is a tick next to OVERWRITE THE OUTPUTS OF GEOPROCESSING OPERATIONS and that there is no tick next to ENABLE under BACKGROUND PROCESSING.

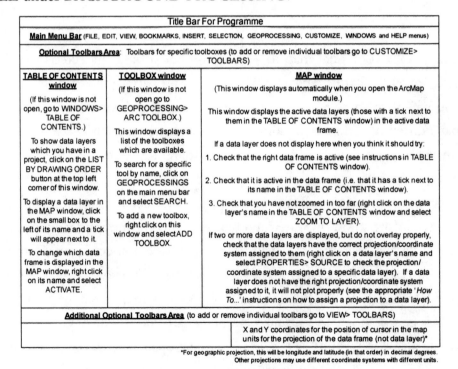

Figure 1. *Schematic of the typical layout of the main ArcGIS user interface (known as ArcMap). Within this book, the individual sections of the window will be referred to using the name which is <u>underlined</u> and in **bold**.*

Chapter Three

Exercise One: Creating A Map Of Species Distribution For A Publication

One of the first, and most common, tasks you will want to do using GIS is to produce maps for use in presentations, reports and publications. This is not as straight-forward as it first may seem and it may take you several attempts to get a map which looks exactly how you want it to. However, it is well worth the effort spent getting it right as a good map is extremely useful for getting information across to other people in an easy to understand format.

You will usually want your final map to have a number of key characteristics. Firstly, you will want it to display all the data layers you wish to show (and no other data layers). Secondly, you will want these to be displayed in a meaningful way. For example, you may want it to have different symbols marking the locations where each species was recorded in a specific study area so that you can tell them apart and compare the distributions of different species. Thirdly, you will want to provide additional information which helps put your data layers in context. For example, in a map of species locational records, you might also want to have information on water depth and the position of any land in the vicinity to make it easier to interpret how the distributions of different species vary in terms of their proximity to the coast or their preferred water depths. Fourthly, you will need to have some information which tells anyone who looks at the map what part of the world it represents. This can be done by putting a 'graticule' around it which show the latitude and longitude for your map around its edges. Finally, you will need to ensure that the projection used for your map is appropriate to the part of the world it represents and does not distort the shape or position of important features within it. This will ensure that the information presented in it can be easily understood. You also want your map to have an appropriate extent. That is, you do not want your map to be so 'zoomed in' that you cannot see all your features of interest, or so 'zoomed out' that the features you are interested in only occupy a very small portion of it and cannot be separated from each other.

You may also want to include a key which tells people what each symbol present on your map means. However, this is optional and you may choose to simply provide a description of the symbols and what they mean in the figure legend (for example, Red circles: Species 1; Blue circles: Species 2; etc). Other optional features include a scale bar so that people know how big features are in your map and an arrow which indicates the direction of North.

As with almost everything in GIS, thinking about what you need to include in your map, and how you are going to display it, before you start will benefit you in the long run. This is because if you rush in without thinking it through properly first, you will undoubtedly make mistakes. Of course, there is nothing wrong with making mistakes, and sometimes it is the best way to learn how to do something, but a bit of thought in advance will save time and effort later.

In this exercise, you will make a map of the distribution of bottlenose dolphin records from surveys in the northern North Sea. Your final map will also include information on water depth and the position of land, so you can get an idea of how bottlenose dolphin are distributed in relation to these two features. While the locational records for bottlenose dolphin provided here are not real, they do represent an approximation of their real distribution in this region.

Before you start this exercise, you will first need to create a new folder on your C: drive called GIS_EXERCISES. To do this on a computer with the Windows 7 operating system, click on the START menu and select MY COMPUTER. In the window which opens, double-click on the icon for your C: drive (this may be called WINDOWS (C:). This will open a window which displays the contents of your C: drive. To create a new folder, right click on this window and select NEW> FOLDER. This will create a new folder. Now call this folder GIS_EXERCISES by typing this into the folder name to replace what it is currently called (which will most likely be NEW FOLDER. This folder, which has the address C:\GIS_EXERCISES\ will be used to store all files and data for the exercises in this book.

Next, you need to download the source files for three data sets from www.gisinecology.com/books/data/marinebiologysupplementaryworkbook. When you download these files, save them into the folder C:\GIS_EXERCISES\ which you have just created. These data sets are:

1. Bottlenose_Dolphin.xls: This is a spreadsheet file which has latitude and longitude for the positions of bottlenose dolphin sightings from a survey conducted in northeast Scotland. **NOTE**: The latitude and longitude have already been converted into decimal degrees. If you are doing this for your own data set, you would need to make sure that all your latitudes and longitudes values have been correctly converted into decimal degrees before you start.

2. Depth_North_Sea: This is a line data layer which contains information on water depth which you can use to create your map. You will need to download all the files called Depth_North_Sea (each with a different extension) in order to be able to use it. This data layer is in the geographic projection and uses the WGS 1984 datum. **NOTE**: The depth values are not accurate and are only provided here as an approximation for illustration purposes. As a result, it should not be used for any purpose other than these exercises.

3. Land_North_Sea: This is a polygon data layer which contains information on land in this region which you can use to create your map. You will need to download all the files called Land_North_Sea (each with a different extension) in order to be able to use it. This data layer is in the geographic projection and uses the WGS 1984 datum. **NOTE**: The land information is not accurate and is only provided here as an approximation for illustration purposes. As a result, it should not be used for any purpose other than these exercises.

Once you have all these files downloaded into the correct folder on your computer, and understand what is contained within each file, you can move onto creating your map. The starting point for this is a blank GIS project. To create a blank GIS project, first, start the ArcGIS software by opening the ArcMap module. When it opens, you will be presented with a window which has the heading ARCMAP – GETTING STARTED. In this window you can either select an existing GIS project to work on, or create a new one. To create a new blank GIS project, click on NEW MAPS in the directory tree on the left hand side and then select BLANK MAP in the right hand section of the window. Now, click OPEN at the bottom of this window. This will open a new blank GIS project. (**NOTE**: If this window does not appear, you can start a new project by clicking on FILE from the main menu bar area, and selecting NEW. When the NEW DOCUMENT window opens, select NEW MAPS and then BLANK MAP in the order outlined above).

Once you have opened your new GIS project, the first thing you need to do is save it under a new, and meaningful, name. To do this, click on FILE from the main menu bar area, and select SAVE AS. Save it as EXERCISE_ONE in the folder C:\GIS_EXERCISES\. The next step is to read over the summary flow diagram for this exercise. This will help you identify all major steps you need to carry out and the individual instruction sets you will need to follow to do them. The summary flow diagram for this exercise is provided on the next two pages.

Summary Flow Diagram For Creating A Map Of Species Distribution

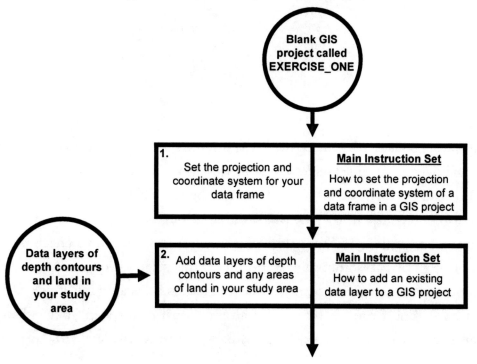

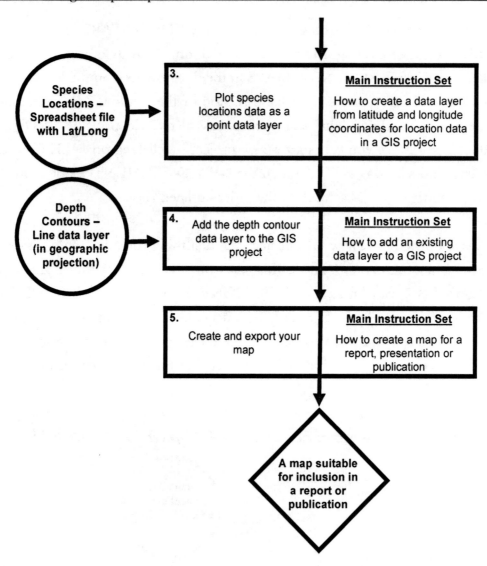

Once you have familiarised yourself with the summary flow diagram outlining the major steps you will need to complete and which instruction sets can be used to complete them, you need to read through the instruction set for the first of these major steps in the flow diagram in its entirety (see below), before working through it stage by stage. Once you have completed the first major step, read through the instructions again to check that you have completed it properly. It is important to do this at this stage as you need to use the results of one major step as the starting point for the next. This is because it is much easier to spot where you have gone wrong at the end of an individual major step, rather than trying to work it out later when you get stuck as a result of a mistake made at this stage. Once you have completed the first major step, move onto the second major step and repeat this process, and so on until you have completed all the steps in the summary flow diagram.

At various points throughout the exercise, images of the contents of the MAP window, the LAYOUT window, the TABLE OF CONTENTS window and/or the TABLE window will be provided so that you have an idea of what your project should look like at that specific point. Check your GIS project against these images. If they do not match up, you will need to go back and work out where you have gone wrong.

NOTE: It is important that you save your GIS project after completing each major step. This can be done by going to the FILE menu on the main menu bar and selecting SAVE.

Instruction Sets For The Individual Steps Identified In The Summary Flow Diagram:

STEP 1: SET THE PROJECTION AND COORDINATE SYSTEM OF YOUR DATA FRAME:

These instructions are based on the instruction set *How to set the projection and coordinate system for a data frame in a GIS project* (from *An Introduction To Using GIS In Marine Biology*). For this exercise, the projection and coordinate system will be set to a custom transverse mercator centred on the middle of the area of data collection. This central position is latitude 56.5°N and longitude 1.0°W (or latitude 56.5° and longitude -1.0° in decimal degrees). This projection was chosen as it provides an accurate representation of the shape and relative positions of features in the study area. In addition, it can also be used to accurately measure distances and areas in real world units, which will be important for later exercises. This custom transverse mercator projection will use the WGS 1984 datum. This datum is used because it is the same as the one used during data collection and for the existing data layers.

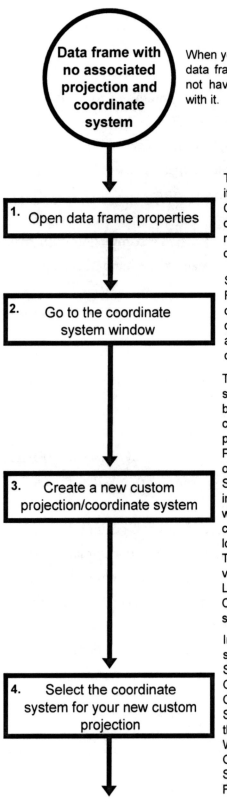

When you open a new GIS project, there will be an empty data frame (called, rather confusingly, LAYERS). It will not have a projection or coordinate system associated with it.

To open the properties of a data frame, right click on its name (in this case LAYERS) in the TABLE OF CONTENTS window, and select the PROPERTIES option. If this is not visible, go to the WINDOWS menus and click on the TABLE OF CONTENTS option.

Select the COORDINATE SYSTEM tab in the DATA FRAME PROPERTIES window. The lower section of this window will tell you which projection and coordinate system are set for the data frame you are currently using. This should currently say 'No coordinate system'.

To create a new custom projection/coordinate system, press the ADD COORDINATE SYSTEM button that can be found towards the top right hand corner of the COORDINATE SYSTEM tab (it has a picture of a globe on it). and select NEW> PROJECTED COORDINATE SYSTEM. This will open the NEW PROJECTED COORDINATE SYSTEM window. In the upper NAME window, type in NORTH SEA. In the PROJECTION portion of the window, select the name of the appropriate type of coordinate system from the drop down menu (in the lower NAME window). For this exercise, select TRANSVERSE MERCATOR. Next type in the values you wish to use for the parameters. LATITUDE_OF_ORIGIN enter 56.5. For CENTRAL_MERIDIAN enter -1.0. Leave all other sections of the window with their default settings.

In the GEOGRAPHIC COORDINATE SYSTEM section of the PROJECTED COORDINATE SYSTEM window, by default it should say NAME: GCS_WGS_1984. If is doesn't, click on the CHANGE button and type WGS 1984 into the SEARCH box in the window that appears and press the return key on your keyboard. Select WORLD> WGS 1984, and click the OK button. Now click the OK button in the NEW PROJECTED COORDINATE SYSTEM window. Finally, click OK in the DATA FRAME PROPERTIES window.

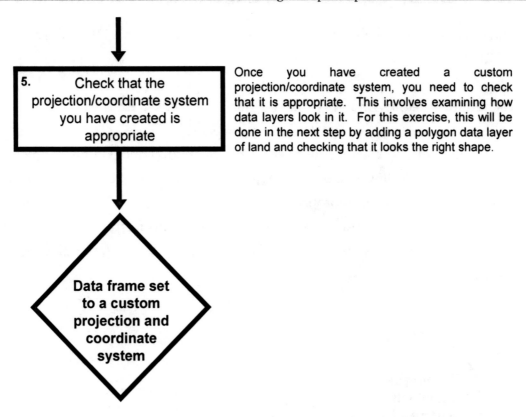

5. Check that the projection/coordinate system you have created is appropriate

Once you have created a custom projection/coordinate system, you need to check that it is appropriate. This involves examining how data layers look in it. For this exercise, this will be done in the next step by adding a polygon data layer of land and checking that it looks the right shape.

Data frame set to a custom projection and coordinate system

To check that you have done this step properly, right click on the name of your data frame (LAYERS) in the TABLE OF CONTENTS window and select PROPERTIES. Click on the COORDINATE SYSTEM tab of the DATA FRAME PROPERTIES window and make sure that the contents of the CURRENT COORDINATE SYSTEM section of the window has the following text at the top of it:

> North Sea
> Authority: Custom
> Projection: Transverse_Mercator
> False_Easting: 0.0
> False_Northing: 0.0
> Central_Meridian: -1.0
> Scale_Factor: 1.0
> Latitude_Of_Origin: 56.5
> Linear Unit: Meter (1.0)

If it does not, you will need to repeat this step until you have assigned the correct projection/coordinate system to your data frame.

STEP 2: ADD A POLYGON DATA LAYER OF ANY AREAS OF LAND IN THE STUDY AREA.

It is usually useful to show any areas of land within your study area. This will help anyone looking at the map work out exactly what part of the world is represented by your map. It will also help you work out whether any other data you add to your GIS project plot in the right place. Finally, it will allow you to assess whether the projection/coordinate system you are using for your data frame is appropriate as you can use it to get an idea of whether any features in the data frame have undue levels of distortion. This instruction set is based on one called *How to add an existing data layer to a GIS project* (from *An Introduction To Using GIS In Marine Biology*).

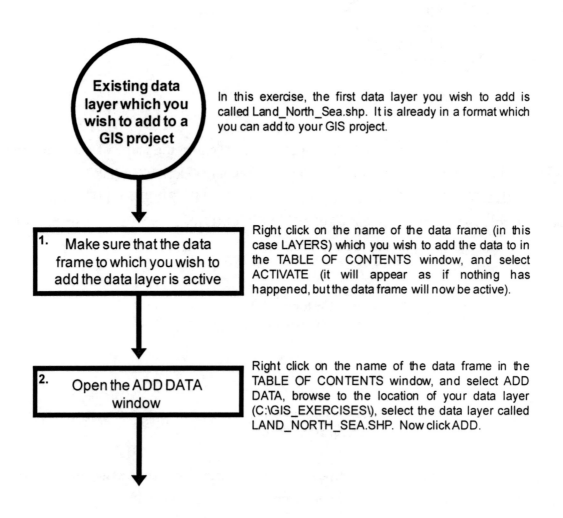

Existing data layer which you wish to add to a GIS project

In this exercise, the first data layer you wish to add is called Land_North_Sea.shp. It is already in a format which you can add to your GIS project.

1. Make sure that the data frame to which you wish to add the data layer is active

Right click on the name of the data frame (in this case LAYERS) which you wish to add the data to in the TABLE OF CONTENTS window, and select ACTIVATE (it will appear as if nothing has happened, but the data frame will now be active).

2. Open the ADD DATA window

Right click on the name of the data frame in the TABLE OF CONTENTS window, and select ADD DATA, browse to the location of your data layer (C:\GIS_EXERCISES\), select the data layer called LAND_NORTH_SEA.SHP. Now click ADD.

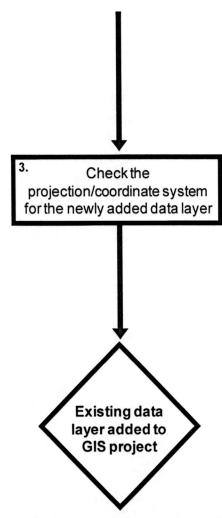

3. Check the projection/coordinate system for the newly added data layer

Existing data layer added to GIS project

Whenever you add a data layer to a GIS project, you should always check that is has a projection/coordinate system assigned to it, and look at what this projection/coordinate system is. This is so that you know whether you will need to assign a projection/coordinate system to it, or transform it into a different projection/coordinate system before you can use it in your GIS project. To check the projection/coordinate system of your newly added data layer. In the TOOLBOX window, go to DATA MANAGEMENT TOOLS> PROJECTIONS AND TRANSFORMATIONS> DEFINE PROJECTION. Select LAND_NORTH_SEA from the drop down menu in the top window. Check its projection/coordinate system in the lower window. This should be GCS_WGS_1984. **NOTE**: If you wish to find out more information about the specific projection/coordinate system of a data layer, you can click on the button at the end of the lower window to display its full details. Once you have looked at these details, click CANCEL to close the SPATIAL REFERENCES PROPERTIES window. Finally, click on the CANCEL button to close the DEFINE PROJECTION window.

At the end of this step, the contents of your MAP window should look like this:

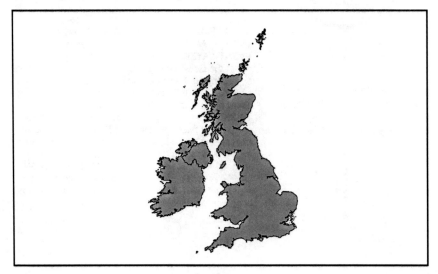

If it does not look like this, try right clicking on the name of the LAND_NORTH_SEA data layer in the TABLE OF CONTENTS window and select ZOOM TO LAYER. If it still does not look right, remove the data layer LAND_NORTH_SEA from your GIS project by right-clicking on its name in the TABLE OF CONTENTS window and selecting REMOVE. Next, go back to step one and ensure that you have set the projection/coordinate system properly and then repeat step two.

STEP 3: PLOT SPECIES LOCATIONS DATA AS A POINT DATA LAYER:

This will use the instruction set *How to create a data layer from latitude and longitude coordinates for locational data in a GIS project* (from *An Introduction To Using GIS In Marine Biology*). This step uses the file Bottlenose_Dolphin.xls and assumes that you have saved it with the following address: C:\GIS_EXERCISES\BOTTLENOSE_DOLPHIN.XLS. It will create a point data layer called BOTTLENOSE_DOLPHIN which will be stored as a shapefile in the folder C:\GIS_EXERCISES\.

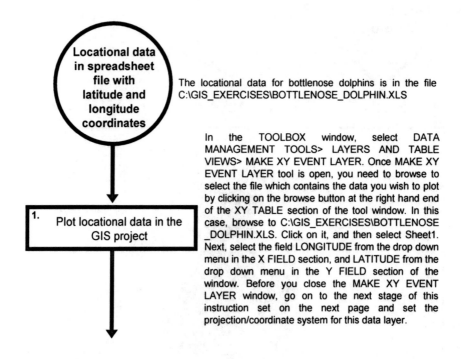

Locational data in spreadsheet file with latitude and longitude coordinates

The locational data for bottlenose dolphins is in the file C:\GIS_EXERCISES\BOTTLENOSE_DOLPHIN.XLS

1. Plot locational data in the GIS project

In the TOOLBOX window, select DATA MANAGEMENT TOOLS> LAYERS AND TABLE VIEWS> MAKE XY EVENT LAYER. Once MAKE XY EVENT LAYER tool is open, you need to browse to select the file which contains the data you wish to plot by clicking on the browse button at the right hand end of the XY TABLE section of the tool window. In this case, browse to C:\GIS_EXERCISES\BOTTLENOSE _DOLPHIN.XLS. Click on it, and then select Sheet1. Next, select the field LONGITUDE from the drop down menu in the X FIELD section, and LATITUDE from the drop down menu in the Y FIELD section of the window. Before you close the MAKE XY EVENT LAYER window, go on to the next stage of this instruction set on the next page and set the projection/coordinate system for this data layer.

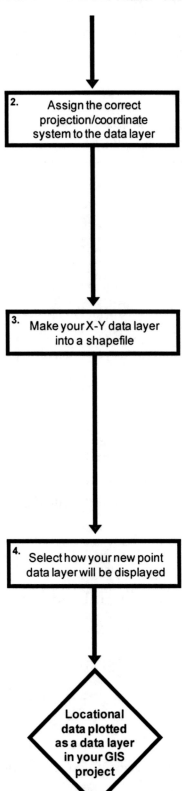

2. Assign the correct projection/coordinate system to the data layer

In the MAKE XY EVENT LAYER window, click on the button at the right hand end of the SPATIAL REFERENCES (OPTIONAL) section. This will open the SPATIAL REFERENCE PROPERTIES window. In this window select GEOGRAPHIC COORDINATE SYSTEMS> WORLD> WGS 1984, then dick on the OK button. You are selecting this projection/coordinate system because you are using latitude and longitude values with the WGS 1984 datum to plot your species locations. **NOTE**: If your coordinates were in a different projection/coordinate system, you would set this different system here. Next, click on OK in the SPATIAL REFERENCE PROPERTIES window. The SPATIAL REFERENCES (OPTIONAL) section should now contain the text GCS_WGS_1984. Finally, click on OK to close the MAKE XY EVENT LAYER window.

3. Make your X-Y data layer into a shapefile

In order to make a permanent version of your data layer, you need to convert it into a shapefile. This can be done using the EXPORT DATA tool. To access this tool, right click on the name of the data layer you just created in the TABLE OF CONTENTS window, and select DATA> EXPORT DATA. This opens the EXPORT DATA window. In the OUTPUT FEATURE CLASS section of this window, type C:\GIS_EXERCISES\BOTTLENOSE_DOLPHIN.SHP and then dick OK. When asked if you want to add the exported data to the map as a layer, click YES. Finally, right click on the name of the data layer you created in stages 1 and 2 (SHEETS$_LAYER) in the TABLE OF CONTENTS window and select REMOVE from the menu which appears to remove it from your GIS project as you no longer need it now you have made it into a shapefile.

4. Select how your new point data layer will be displayed

Right click on the name of the BOTTLENOSE_DOLPHIN data layer in the TABLE OF CONTENTS window, and select PROPERTIES. In the LAYER PROPERTIES window, click on the SYMBOLOGY tab. On the left-hand side, under SHOW select FEATURES> SINGLE SYMBOL. Next, under SYMBOL in the top middle of the LAYER PROPERTIES window, click on the button with the coloured circle on it. This will open the SYMBOL SELECTOR window. Select CIRCLE 2 for the symbol by clicking on it in the left hand section of the window. Next, on the right hand side of the SYMBOL SELECTOR window, click on the button next to COLOUR and hover the cursor over the fourth red box down. This will tell you that this colour's name is PONSETTIA RED. Select this by dicking on it. Next, select 12.0 for size. Finally, click OK to close the SYMBOL SELECTOR window and then OK to close the LAYER PROPERTIES window.

Locational data plotted as a data layer in your GIS project

Once you have completed this instruction set, right-click on the data layer name BOTTLENOSE_DOLPHIN in the TABLE OF CONTENTS window and select ZOOM TO LAYER.

The contents of your MAP window, should now look like this:

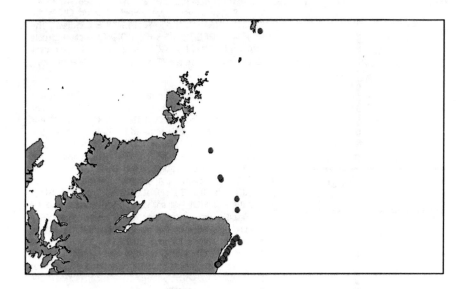

If it does not look like this, ensure that you set the symbology settings exactly as detailed in the instructions. After this, if you are still having problems, check that you used longitude as the X coordinate and latitude as the Y coordinate when plotting this data layer. Finally, check that you assigned the correct projection/coordinate system to the BOTTLENOSE_DOLPHIN data layer. To do this, right click on its name in the TABLE OF CONTENTS window and select PROPERTIES. Click on the SOURCE tab and ensure that at the bottom of the DATA SOURCE section of the window you have the following text:

> Geographic Coordinate System: GCS_WGS_1984
> Datum: D_WGS_1984
> Prime Meridian: Greenwich
> Angular Unit: Degree

If it does not, you will need to repeat step three from the beginning, ensuring that you set the projection/coordinate system to the correct one given in the above instruction set.

STEP 4: ADD THE DEPTH CONTOUR DATA LAYER TO THE GIS PROJECT:

This will use the instruction set called *How to add an existing data layer to a GIS project* (from *An Introduction To Using GIS In Marine Biology*), and is essentially the same as step two of this exercise. **NOTE**: This data layer already has a projection and coordinate system assigned to it. This is a geographic projection based on the WGS 1984 datum.

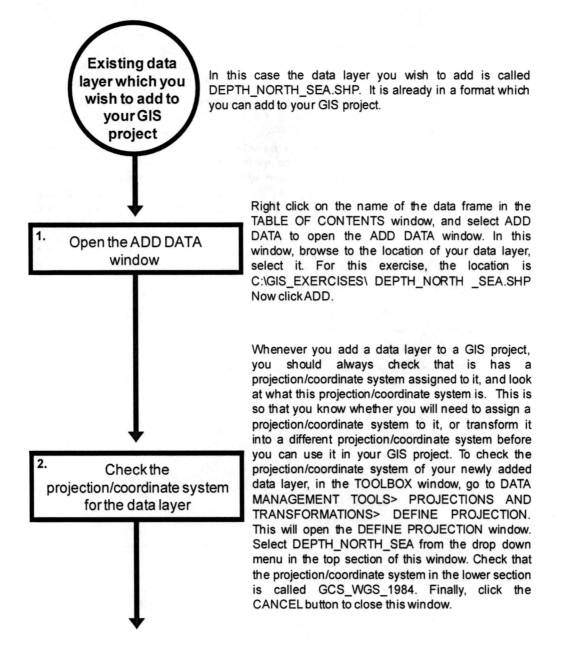

Existing data layer which you wish to add to your GIS project

In this case the data layer you wish to add is called DEPTH_NORTH_SEA.SHP. It is already in a format which you can add to your GIS project.

1. Open the ADD DATA window

Right click on the name of the data frame in the TABLE OF CONTENTS window, and select ADD DATA to open the ADD DATA window. In this window, browse to the location of your data layer, select it. For this exercise, the location is C:\GIS_EXERCISES\ DEPTH_NORTH _SEA.SHP Now click ADD.

2. Check the projection/coordinate system for the data layer

Whenever you add a data layer to a GIS project, you should always check that is has a projection/coordinate system assigned to it, and look at what this projection/coordinate system is. This is so that you know whether you will need to assign a projection/coordinate system to it, or transform it into a different projection/coordinate system before you can use it in your GIS project. To check the projection/coordinate system of your newly added data layer, in the TOOLBOX window, go to DATA MANAGEMENT TOOLS> PROJECTIONS AND TRANSFORMATIONS> DEFINE PROJECTION. This will open the DEFINE PROJECTION window. Select DEPTH_NORTH_SEA from the drop down menu in the top section of this window. Check that the projection/coordinate system in the lower section is called GCS_WGS_1984. Finally, click the CANCEL button to close this window.

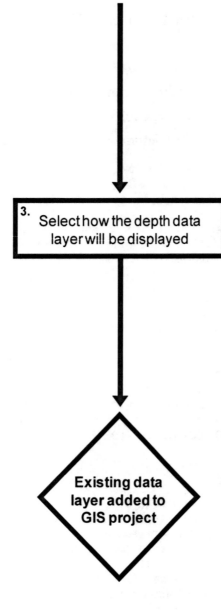

3. Select how the depth data layer will be displayed

Existing data layer added to GIS project

Right click on the name of the DEPTH_NORTH_SEA shapefile data layer in the TABLE OF CONTENTS window, and select PROPERTIES. In the LAYER PROPERTIES window, click on the SYMBOLOGY tab. On the left-hand side under show, select CATEGORIES> UNIQUE VALUES. Next, under VALUE FIELD select CONTOUR. Next, click on the ADD VALUES button towards the bottom middle of the LAYER PROPERTIES window to open the ADD VALUES window. From the list of values, scroll down until you find -10, and click on it. Hold the CTRL button down on your keyboard and select -20, -50, -100 and -150 as well. Now click OK.

You can now select colours for these contours in the LAYER PROPERTIES window. In the LAYER PROPERTIES window under SYMBOL, click on the box to the left of <ALL OTHER VALUES> to remove the tick inside it. This means that only the contours you have selected will be displayed. Now, double-click on the horizontal line next to each contour value and select a different shade of blue for each one, going from a light shade for the shallowest depth to a dark shade for the deepest depth. Once you have done this, click OK to close the LAYER PROPERTIES window.

When you have finished step four, the TABLE OF CONTENTS window should look like one of the two images below (depending on whether the DEPTH_NORTH_SEA data layer was added above of below the BOTTLENOSE DOLPHIN data layer):

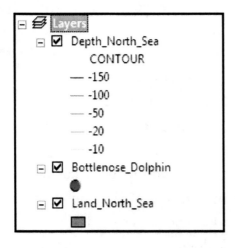 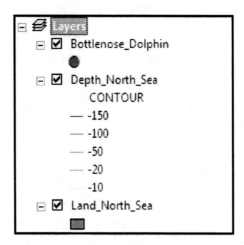

If you have a picture of a folder and a file directory address above each data layer name in the TABLE OF CONTENTS window, this is OK. These can be hidden by clicking on the LIST BY DRAWING ORDER button at the top left hand corner of the TABLE OF CONTENTS window.

The contents of your MAP window should look like this:

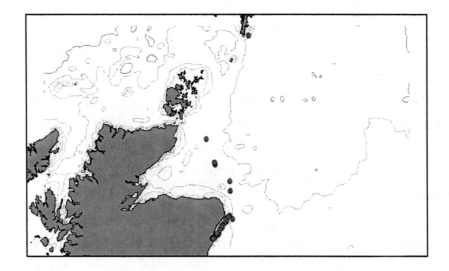

STEP 5: CREATE AND EXPORT YOUR MAP:

Now that you have all the required data added to your GIS project, you can get an idea of how your final map will look. It is unlikely that you will get your map looking perfect on the first attempt, and you will almost always have to change something. The instructions for this step are based on the instruction set called *How to create a map for a report, presentation or publication* (from *An Introduction To Using GIS In Marine Biology*). They are relatively simple and so are provided as text without an accompanying flow diagram. Images are provided along the way to allow you to check that you have completed each section correctly.

Open the LAYOUT window by going to the VIEW menu on the main menu bar and selecting LAYOUT VIEW. (If you want to get back to the MAP window at any point, simply go to the VIEW menu on the main menu bar and select DATA VIEW.) In the LAYOUT window (which will replace the MAP window), you will see what your map will look like for your specific data frame. When you initially look at it, you will undoubtedly not be too impressed. However, with a few simple steps, you can make it look much better.

The first thing that you will need to do is to change the extent of the data frame so that your map only shows the area you want it to. To do this, right-click on the name of your data frame (LAYERS) in the TABLE OF CONTENTS window and select PROPERTIES. In the DATA FRAME PROPERTIES window, select the DATA FRAME tab. Here, you will find the EXTENT options. To limit the extent of your map, select FIXED EXTENT from the drop down menu. You can then enter the required coordinates for the limits of the extent by filling in the options that appear. In this exercise, you want to focus in on the areas where bottlenose dolphins were recorded. This can be done by setting the TOP limit to 388000, the LEFT limit to -130000, the RIGHT limit to 50000, and the BOTTOM limit to 56000. Click the APPLY button at the bottom of the DATA FRAME PROPERTIES window and you will see the extent of your map will change in the LAYOUT window.

NOTE: If you did not already have these coordinates, you could get an idea of what coordinates would be appropriate to define the extent for a specific map before you open the DATA FRAME PROPERTIES window, by going to the MAP window and ZOOM

IN or ZOOM OUT until you can see all the data you wish to show on your map. Next, move the cursor to the top left of the area you wish your map to show. You can then read off the coordinates which you would need to use to set the left (the first coordinate) and top (the second coordinate) limits for the extent of your map by looking at the coordinate display area at the bottom right-hand corner of the ArcMap user interface (see figure 1). Repeat this for the bottom right-hand corner of the area you wish your map to show and read off the right (the first coordinate) and bottom (the second coordinate) limits for the extent of you map. Once you have these values, you can set the extent of your map as outlined above. If you want to give this a go, you would first need to change the extent setting from FIXED EXTENT to AUTOMATIC to allow you to zoom in and out in the MAP window. However, remember to set the extent back to a fixed extent using the above coordinates before you continue with this exercise. If you do not, the contents of the LAYOUT window and the final map will not match the images provided.

Next, you want to set the size and the position of the window. Select the SIZE AND POSITION tab of the DATA FRAME PROPERTIES window. For POSITION, enter 5cm for X and 6cm for Y and click APPLY, you will see the position of your map change. Finally, for size enter 11cm for WIDTH and then click on the HEIGHT box. This will automatically update the HEIGHT value to around 20cm to maintain the aspect ratio of the data frame (which is set by the extent coordinates you entered above). You can now click OK to close the DATA FRAME PROPERTIES window and have a closer look at your map in the LAYOUT window. It should now look like the image at the top of the next page.

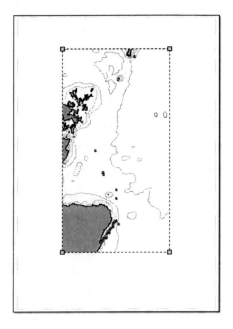

If it does not look like this, go back and check that you set the extent, size and position to the correct values given in the instructions before carrying on.

The next thing you need to do is to check the order of data layers in your project to ensure that they overlay each other correctly. If you look at your TABLE OF CONTENTS window, you may see that the DEPTH_NORTH_SEA data layer is on the top, followed by BOTTLENOSE_DOLPHIN and LAND_NORTH_SEA (depending on where it appeared when you first added the DEPTH_NORTH_SEA data layer). If this is the case, you will see that this means that some of the depth contours are on top of some of the bottlenose dolphin locations. This looks messy, and it would be better if the bottlenose dolphin locations were on the top. This can be sorted by clicking on the DEPTH_NORTH_SEA data layer in the TABLE OF CONTENTS window and holding the left hand mouse button down. You can then drag it downwards to change the order and place it below the BOTTLENOSE_DOLPHIN data layer.

This will change the order of data layers in your TABLE OF CONTENTS window as shown at the top of the next page.

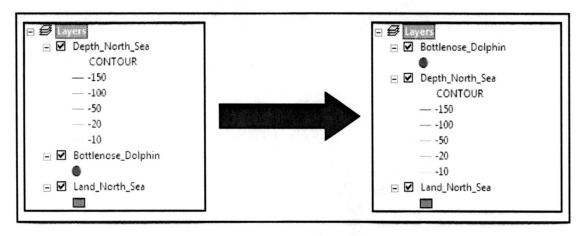

The next thing to do is to add a latitude and longitude grid around the edges of your map so that people will know what part of the world it represents. As with the extent, size and position, this is also done in the DATA FRAME PROPERTIES window. Open it again by right-clicking on the name of your data frame (LAYERS) in the TABLE OF CONTENTS window and selecting PROPERTIES. This time click on the GRIDS tab. Now click on the NEW GRID button. This will open the GRIDS AND GRATICULES wizard, and you can work your way through the options to get your latitude and longitude grid looking exactly the way that you want. For the purposes of this exercise, you will mostly use the default settings for your latitude and longitude grid, but there are many options which you can use to customise how your latitude and longitude grid will look. Once you have a bit more experience you can play around with all the options until you find one you like.

Firstly, in the GRIDS AND GRATICULES wizard, you will select GRATICULES: DIVIDES MAP BY MERIDIANS AND PARALLELS and then click the NEXT button. This will take you to the CREATE A GRATICULE window. Select LABELS ONLY and set the INTERVALS for both parallels and meridians to 1 degree. Then click the NEXT button. This will take you to the AXES AND LABELS window. Click on the TEXT STYLE button and in the SYMBOL SELECTOR window which opens set the size to 16 and click OK. Click the NEXT button in the AXES AND LABELS window. Click the FINISH button to close the AXES AND LABELS window. Finally click OK to close the DATA FRAME PROPERTIES window.

The contents of your LAYOUT window should now look like this:

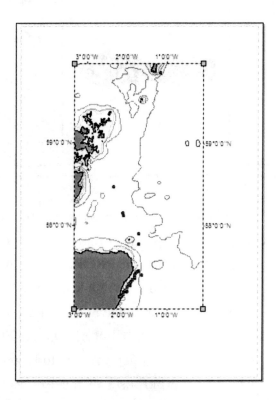

Next, you might want to add additional information, such as a scale bar, a north arrow or a legend. These can be added using the options under the INSERT menu on the main menu bar.

For this exercise, you will insert a scale bar. To do this, click on the INSERT menu on the main menu bar and select SCALE BAR. This will open the SCALE BAR SELECTOR window. You will find there are many different ways you can set your scale bar, but for this exercise, you will use a relatively simple one. Click on the top scale bar design option on the left hand side of the SCALE BAR SELECTOR window. Then click on the PROPERTIES button to open the SCALE BAR window. Here you can change the settings to get exactly the right design of scale bar you wish. For this exercise, you will simply use all the default options, with the exception of the DIVISION UNITS option. You will change this from MILES to KILOMETRES. Click OK to close the SCALE BAR window, and then OK to close the SCALE BAR SELECTOR window. You will see a scale bar has now been added to your map in the LAYOUT window. The first thing to do is to change its size if required. This is done by double clicking on the scale bar to open the SCALE LINE PROPERTIES window and selecting the SIZE AND POSITION tab. For this exercise, enter a value of 6cm for the width and then click OK. This will set the scale bar to a reasonable length for your map. Finally, you need to position the scale bar where you want it to be. For this exercise, you will want to move the scale bar down to the lower right hand corner of the map. This is done by clicking on it and then dragging it while holding the left hand mouse button down.

The contents of your LAYOUT window should now look like this:

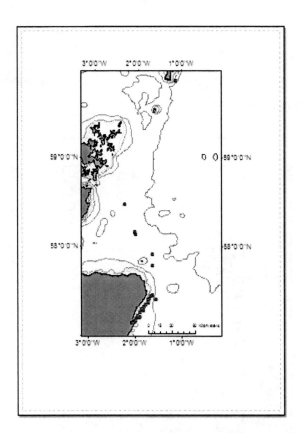

Since this is a relatively simple map, you can describe its contents in a figure legend which will tell the viewer what the different symbols and colours mean. For more complex maps, you might want to add a specific legend. This can be done through the INSERT menu.

All that is left now is for you to export your map using the EXPORT MAP tool. This is selected by clicking on the FILE menu on the main menu bar and selecting EXPORT MAP. This will open the EXPORT MAP window where you can select the format and resolution you wish to export the map in and the location where you wish to save it. The format and the resolution you select will depend on what you wish to use your map for. For example, you may choose a different format and resolution depending on whether you are creating a map to include in a presentation, or one for a written report. For this exercise, export your map as a jpg with a resolution of 300 dpi, which would be suitable for most reports and manuscripts, call it BND_MAP_EXERCISE_ONE and save it in the folder C:\GIS_EXERCISES\.

Optional Extra:

If you wish to further explore creating maps of species distribution for a publication, you can try to create the following figure:

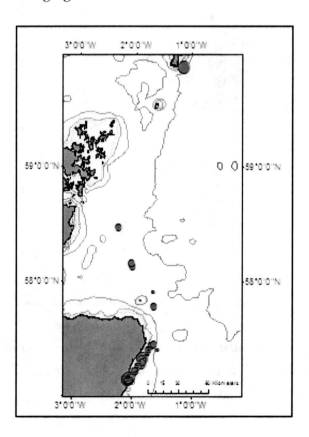

This figure differs from the original one created in this exercise because it uses symbols of different sizes to tell you not only where bottlenose dolphin were recorded, but also how big each group was. It was created by changing the way that the bottlenose dolphin locations are displayed.

To make this map, right click on the data layer name BOTTLENOSE_ DOLPHIN in the TABLE OF CONTENTS window and select PROPERTIES. Click on the SYMBOLOGY tab. However, rather than selecting FEATURES> SINGLE SYMBOL under SHOW on the left hand side of the window, select QUANTITIES> GRADUATED SYMBOLS instead. Next, click on the FIELD window and select NUMBER. You will now see in the lower section of the window, five ranges of group size have been created automatically. However, you will want to customise this. To do this, first click on RANGE for the smallest symbol and type in the number 1. For the next symbol down, type in 5, then 10 for the third and 25 for the fourth one. Lastly, right click on the fifth one, and select REMOVE CLASS(ES). This will leave you with four categories covering the following group sizes: 1, 2-5, 6-10 and 11-25. Next, click on the TEMPLATE button and select CIRCLE 2 in the SYMBOL SELECTOR window when it opens. Now click on COLOUR and select PONSETTIA RED as the colour and then click on the OK button. Finally, in the SYMBOLOGY window, in the SYMBOL SIZE FROM section in the middle of the window type the number 10 and in the TO section beside it type 30 before clicking on the OK button. You will now see that the symbols marking the locations of bottlenose dolphin records are different sizes and that the map looks like the one provided above.

Chapter Four

Exercise Two: Creating A Presence-Absence Raster Data Layer For A Species From Survey Data

In exercise one, you produced a simple map of all the locations where one species, the bottlenose dolphin, were recorded during surveys conducted in the seas around northeast Scotland. While such maps are sometimes useful, they do not tell you anything about which areas the surveys covered and which they did not. Therefore, it is difficult to work out whether the absence of the species from a specific area is due to a lack of dolphin presence, or a lack of survey coverage. This means that it would be good to also have information about survey coverage within your GIS project, and on your map. One way of showing which areas were surveyed and which were not in relation to sightings locations is to create what is known as a presence-absence raster data layer (examples of another way in which survey coverage can be taken into account can be found in exercises four and five later in this book). A presence-absence raster data layer is made by dividing the study area up into grid cells and working out whether each grid cell was surveyed and whether a species of interest was recorded in it or not. As well as being useful for displaying the distribution of a species in relation to survey coverage, such raster data layers can also form the basis for studies of habitat preferences and species distribution modelling. Therefore, in this exercise, you will create a presence-absence raster data layer for bottlenose dolphins for the survey data for northeast Scotland.

For this exercise, you will use two data layers. One of these is a data layer which was created in exercise one, so you will need to have successfully completed this exercise first before trying to do this one. These data layers should all be saved in a folder on your C drive called GIS_EXERCISES, so that it has the address C:\GIS_EXERCISES\.

The data layers needed for this exercise are:

1. Bottlenose_Dolphin.shp: This is a point data layer and is the shapefile of bottlenose dolphin locational records created in step two of exercise one. Each point represents a record for a single group of dolphins recorded during a set of surveys. It is in the geographic projection and is based on the WGS 1984 datum.

2. Survey_Tracks_North_Sea.shp: This is a line data layer which contains information on survey effort for the study area which recorded the bottlenose dolphin locational data in Bottlenose_Dolphin.shp. Thus, this data layer tells you which locations were surveyed in this study. It is in the geographic projection and is based on the WGS 1984 datum.

The starting point for this exercise is the GIS project created in exercise one. Open the ArcMap module of the ArcGIS 10.1 software. When it opens, you will be presented with a window which has the heading ARCMAP – GETTING STARTED. To open an existing GIS project, click on EXISTING MAPS in the directory tree on the left hand side and then select the project called EXERCISE_ONE in the right hand section of the window. Now, click OPEN at the bottom of this window. If it is not listed here, click on the BROWSE FOR MORE option on the left had side of the window and use the browser window to find and select this GIS project from the folder C:\GIS_EXERCISES\. Now, click OPEN at the bottom of this window. If the ARCMAP – GETTING STARTED window does not appear when you start the ArcMap module, you can open an existing project by clicking on FILE from the main menu bar area, and selecting OPEN. When the OPEN window opens, browse to the location where your project is saved (C:\GIS_EXERCISES\), select it and click OPEN.

Once you have opened the GIS project called EXERCISE_ONE, the first thing you need to do is save it under a new name. This is because you do not want to alter the contents of the original project, you just want to base your new one on it since this saves you having to add all the data layers again, and also having to reset the projection/coordinate system of the data frame. To save the project under a new name, click on FILE from the main menu bar area, and select SAVE AS. For this example, save it as EXERCISE_TWO in the C:\GIS_EXERCISES\ folder.

This project should have the following layer in it: 1. Bottlenose_Dolphin.shp; 2. Depth_North_Sea.shp; 3. Land_North_Sea. Thus, at the start of this exercise, your MAP window should look like the image below (**NOTE**: You will have to go to the VIEW menu on the main menu bar and select DATA VIEW to get back to the MAP window).

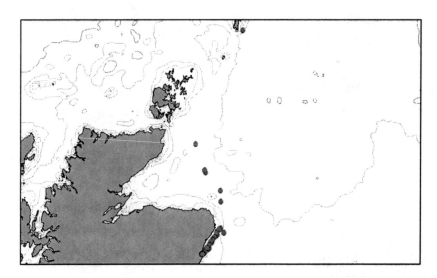

The next step is to read over the summary flow diagram for this exercise, which is provided below.

Summary Flow Diagram For Creating A Presence-Absence Raster Data layer

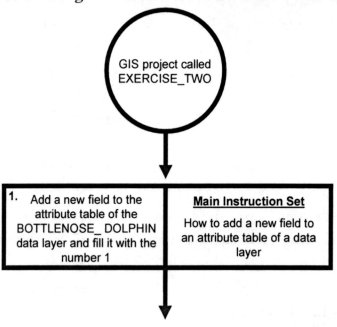

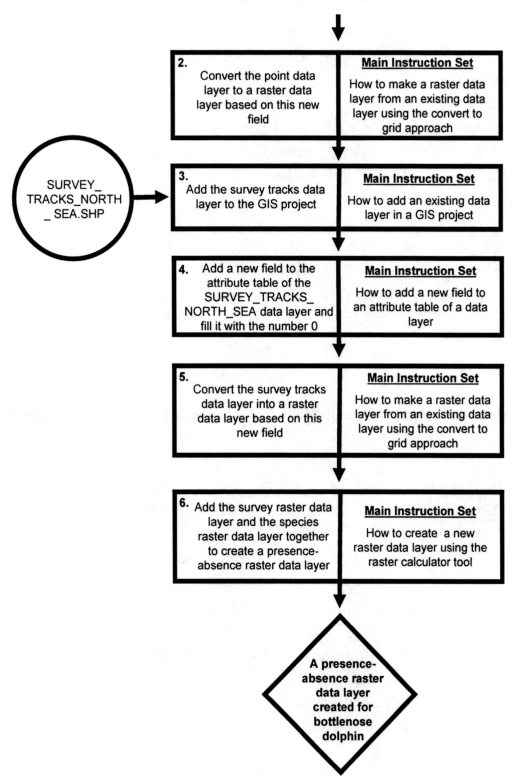

Once you have familiarised yourself with this summary flow diagram, you need to read through the instruction set for the first step in the flow diagram in its entirety, before working through it.

When you have completed the first major step, read through the instructions again and make sure that you have completed it properly. Next, move onto the second major step and repeat this process, and so on until you have completed all the steps in the summary flow diagram.

At various points, images of the contents of the MAP window, the TABLE OF CONTENTS window and/or the TABLE window will be provided so that you have an idea of what your GIS project should look like at specific points as you progress through this exercise.

Instruction Sets For The Individual Steps Identified In The Summary Flow Diagram:

STEP 1: ADD A NEW FIELD TO THE ATTRIBUTE TABLE OF THE BOTTLENOSE DOLPHIN DATA LAYER AND FILL IT WITH THE NUMBER 1:

In order to make a presence-absence raster data layer, you first need to create a presence raster data layer for your species, this will be done in step two. However, before you can do this you need to have a field in the attribute table of your data layer of the species locations (i.e. BOTTLENOSE_DOLPHIN) which has a value of one in it for all records. This indicates species presence and is created in step one. The instruction set for this step is primarily based on one called *How to add a new field to an attribute table*, but it also includes steps from *How to use the field calculator tool to fill in values in a new field*. Generic versions of these instruction sets can be found in *An Introduction To Using GIS In Marine Biology*.

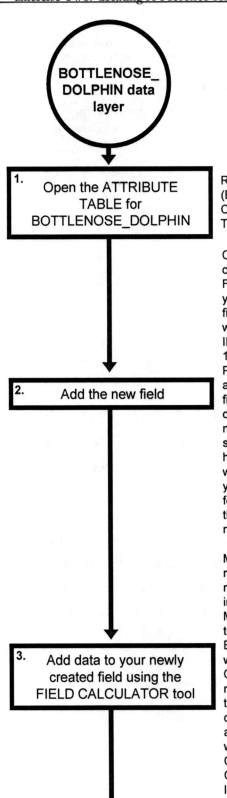

BOTTLENOSE_ DOLPHIN data layer

1. Open the ATTRIBUTE TABLE for BOTTLENOSE_DOLPHIN

Right click on the name of the data layer (BOTTLENOSE_DOLPHIN) in the TABLE OF CONTENTS window and select OPEN ATTRIBUTE TABLE.

Click the TABLE OPTIONS button at the top left corner of the TABLE window and select ADD FIELD. This opens the ADD FIELD window, where you can enter the required settings for your new field. Type the word PRESENCE in the NAME window. In the TYPE window select SHORT INTEGER (since you simply want to add the number 1 to this field rather than use decimal places). In the PRECISION window, enter the number 16 (this allows you to put a maximum of 16 digits in this field). Now click OK. **NOTE**: Once a field has been created, you cannot change these properties or its name so it is important to get this right from the start. If you add the field and then you find that you have got the settings wrong, the best way to deal with this is to delete it and start again. **NOTE**: When you create a field, it will usually have values of zero for all lines in it. You will need to update this to have the value you want in it. This will be done in the next stage.

2. Add the new field

Move the TABLE window so that you can see the main menu bar and optional toolbar areas of the main MAP window. Click on the CUSTOMIZE menu in the main menu bar and select TOOLBARS. Make sure there is a tick next to the EDITOR toolbar. In the EDITOR toolbar, click on the EDITOR button and select START EDITING. If a warning window appears, this is okay. Just Click CONTINUE and carry on. A new EDITOR window may then appear (usually on the right hand side of the main MAP window – do not worry if this window does not appear). Now click on the TABLE window and then right click on the field called PRESENCE which you just created and select FIELD CALCULATOR This will open the FIELD CALCULATOR window. Type the number 1 into the lower window. Click OK. Finally, close the TABLE window.

3. Add data to your newly created field using the FIELD CALCULATOR tool

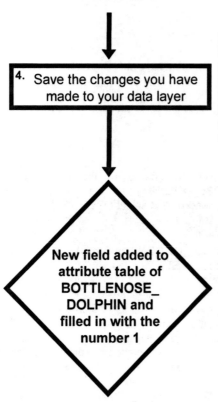

4. Save the changes you have made to your data layer

Now that you have finished editing the attribute table for the BOTTLENOSE_DOLPHIN data layer, you need to save the edits. To do this, in the EDITOR toolbar, click on EDITOR and select SAVE EDITS. Next, in the EDITOR toolbar, click on the EDITOR button and select STOP EDITING. **NOTE**: If you were adding a lot of data, you should save the changes to the table as you go along.

New field added to attribute table of BOTTLENOSE_ DOLPHIN and filled in with the number 1

NOTE: If the EDITOR window (sometimes also labelled as the CREATE FEATURES window) remains open once you have stopped editing the data layer, close it before you carry on with this exercise.

When you have completed this step, the attribute table for the BOTTLENOSE_DOLPHIN data layer should have changed from looking like the top half of the figure at the top the next page to looking like the bottom half, with a new field called Presence added, which has a 1 in it for each line. If the attribute table is not open, right click on the name BOTTLENOSE_DOLPHIN in the TABLE OF CONTENTS window and select OPEN ATTRIBUTE TABLE.

FID	Shape *	Sighting_I	Latitude	Longitude	Species	Number
0	Point	500	57.148333	-2.069667	Bottlenose dolphin	15
1	Point	501	57.1515	-2.045167	Bottlenose dolphin	5
2	Point	502	57.1525	-2.042667	Bottlenose dolphin	5
3	Point	506	57.141667	-2.070833	Bottlenose dolphin	6
4	Point	511	57.148	-2.051667	Bottlenose dolphin	6
5	Point	535	57.151167	-2.045333	Bottlenose dolphin	4
6	Point	554	57.149667	-2.05	Bottlenose dolphin	9
7	Point	591	58.472667	-2.27	Bottlenose dolphin	2
8	Point	592	57.15	-2.034667	Bottlenose dolphin	6
9	Point	595	57.157167	-2.033333	Bottlenose dolphin	6
10	Point	596	57.150833	-2.046833	Bottlenose dolphin	10
11	Point	600	57.146833	-2.054833	Bottlenose dolphin	7
12	Point	601	57.262333	-1.897333	Bottlenose dolphin	1
13	Point	603	57.373167	-1.7365	Bottlenose dolphin	1
14	Point	604	57.783333	-1.636333	Bottlenose dolphin	2
15	Point	632	57.147167	-2.059	Bottlenose dolphin	4
16	Point	657	57.455667	-1.6355	Bottlenose dolphin	2
17	Point	658	57.381333	-1.719667	Bottlenose dolphin	1
18	Point	659	57.381333	-1.719667	Bottlenose dolphin	1
19	Point	711	57.285333	-1.8405	Bottlenose dolphin	6
20	Point	717	57.143	-2.065333	Bottlenose dolphin	15

FID	Shape *	Sighting_I	Latitude	Longitude	Species	Number	Presence
0	Point	500	57.148333	-2.069667	Bottlenose dolphin	15	1
1	Point	501	57.1515	-2.045167	Bottlenose dolphin	5	1
2	Point	502	57.1525	-2.042667	Bottlenose dolphin	5	1
3	Point	506	57.141667	-2.070833	Bottlenose dolphin	6	1
4	Point	511	57.148	-2.051667	Bottlenose dolphin	6	1
5	Point	535	57.151167	-2.045333	Bottlenose dolphin	4	1
6	Point	554	57.149667	-2.05	Bottlenose dolphin	9	1
7	Point	591	58.472667	-2.27	Bottlenose dolphin	2	1
8	Point	592	57.15	-2.034667	Bottlenose dolphin	6	1
9	Point	595	57.157167	-2.033333	Bottlenose dolphin	6	1
10	Point	596	57.150333	-2.046833	Bottlenose dolphin	10	1
11	Point	600	57.146833	-2.054833	Bottlenose dolphin	7	1
12	Point	601	57.262333	-1.897333	Bottlenose dolphin	1	1
13	Point	603	57.373167	-1.7365	Bottlenose dolphin	1	1
14	Point	604	57.783333	-1.636333	Bottlenose dolphin	2	1
15	Point	632	57.147167	-2.059	Bottlenose dolphin	4	1
16	Point	657	57.455667	-1.6355	Bottlenose dolphin	2	1
17	Point	658	57.381333	-1.719667	Bottlenose dolphin	1	1
18	Point	659	57.381333	-1.719667	Bottlenose dolphin	1	1
19	Point	711	57.285333	-1.8405	Bottlenose dolphin	6	1
20	Point	717	57.143	-2.065333	Bottlenose dolphin	15	1

Now close the attribute table for the bottlenose dolphin data layer.

STEP 2: CONVERT BOTTLENOSE DOLPHIN DATA LAYER TO A RASTER DATA LAYER BASED ON THE NEW FIELD CALLED 'PRESENCE':

In this step you will create a raster data layer where the cell values will indicate whether bottlenose dolphin were recorded as present during the study. It will use the values in the PRESENCE field of the attribute table of the BOTTLENOSE_DOLPHIN data layer created in step one to calculate the value for each cell, and when more than one bottlenose dolphin record falls within the same cell, it will calculate an average value for that cell. Since all the bottlenose dolphin records in the BOTTLENOSE_DOLPHIN data layer have a value of one for the presence field, this will result in a cell value of one regardless of how many sightings were recorded in an individual cell. A cell size of 10km (or 10,000m) will be used for the raster data layer's resolution. This cell size was selected as it provides a raster data layer where the cells are not too big or too small for the study area, and which is appropriate to both the distribution of bottlenose dolphins in this region and the spatial variation in environmental variables. The extent will be set to match up with that used for the map created in exercise one. This is because this covers all the areas where surveys were conducted. In terms of a projection/coordinate system, it will use the same one as the data frame. This is because it allows you to create a raster data layer which is measured in real world units (in this case metres) and is appropriate to the size and position of the study area. However, since the BOTTLENOSE_DOLPHIN data layer is not in this projection/coordinate system, it will have to be transformed into it before the presence raster data layer is made. In the raster data layer created in this step, which will be called BND_RASTER, any cells where bottlenose dolphin were recorded will have a value of one, and all other cells will be classified as NO DATA, and will not be displayed.

The instructions for this step (which start on the next page) are based on instruction sets called *How to transform data layers between different projections*, *How to create a new raster data layer from an existing point data layer* and *How to change the way a raster data layer is displayed*. Generic versions of these instruction sets can be found in *An Introduction To Using GIS In Marine Biology*. The generic instruction set for *How To create a new raster data layer from an existing point data layer* includes advice for selecting appropriate cell sizes, extents and projection/coordinate systems for a specific data set and study area. However, for this exercise, you will use the ones detailed above.

NOTE: If you are using ArcGIS 10.2, you may find that the new transformed data layer is not automatically added to your GIS project when you run the PROJECT tool. If this is the case, simply right-click on the data frame's name (LAYERS) in the TABLE OF CONTENTS window and select ADD DATA. Broswe to the location where your newly created data layer is stored, selectin it and click the ADDbutton.

BOTTLENOSE_ DOLPHIN point data layer which you wish to convert into a raster data layer

1. Convert your point data layer to the same projection/coordinate system as the data frame

In the TOOLBOX window, select DATA MANAGEMENT TOOLS> PROJECTIONS AND TRANSFORMATIONS> FEATURE> PROJECT. This will open the PROJECT window. Select BOTTLENOSE_DOLPHIN from the drop down menu in the INPUT DATASET OR FEATURE CLASS window. In the OUTPUT DATASET OR FEATURE CLASS window enter C:\GIS_EXERCISES \BOTTLENOSE_DOLPHIN_PROJECT.

Next, click on the button at the end of the OUTPUT COORDINATE SYSTEM section of the window to open the SPATIAL REFERENCE PROPERTIES window. To create the custom projection/coordinate system being used for this exercise, click on the ADD COORDINATE SYSTEM button towards the top right hand corner of the SPATIAL REFERENCE PROPERTIES window and select NEW> PROJECTED COORDINATE SYSTEM. This will open the NEW PROJECTED COORDINATE SYSTEM window. In the NAME window, type in NORTH SEA. In the PROJECTION portion of the window, select TRANSVERSE MERCATOR from the drop down menu. Next type 56.5 for LATITUDE_OF_ORIGIN and -1.0 for CENTRAL_MERIDIAN.

In the GEOGRAPHIC COORDINATE SYSTEM section of the NEW PROJECTED COORDINATE SYSTEM window, by default it should say NAME: GCS_WGS_1984. If is doesn't, click on the CHANGE button and type WGS 1984 into the SEARCH box in the window that appears and press the return key on your keyboard. Select WORLD> WGS 1984, and click the OK button. Now click the OK button in the NEW PROJECTED COORDINATE SYSTEM window. Next click the OK button in the SPATIAL REFERENCE PROPERTIES window. Finally, click OK in the PROJECT window.

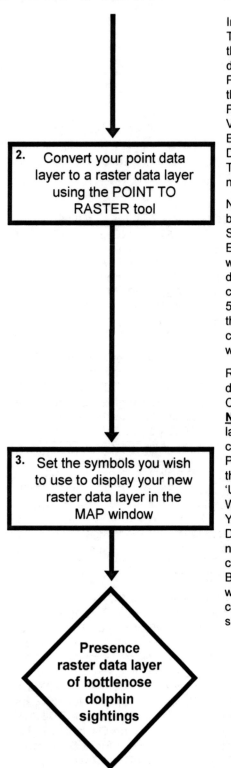

2. Convert your point data layer to a raster data layer using the POINT TO RASTER tool

3. Set the symbols you wish to use to display your new raster data layer in the MAP window

Presence raster data layer of bottlenose dolphin sightings

In the toolbox window, select CONVERSION TOOLS > TO RASTER > POINT TO RASTER. In the POINT TO RASTER window, select the point data layer called BOTTLENOSE_DOLPHIN_ PROJECT in the INPUT FEATURES window using the drop down menu. Select the field called PRESENCE using the drop down menu in the VALUE FIELD window. Type C:\GIS_EXERCISES\ BND_RASTER into the OUTPUT RASTER DATASET window. In the CELL ASSIGNMENT TYPE (OPTIONAL) window select MEAN. Type the number 10000 into CELLSIZE (OPTIONAL) window.

Next, click on the ENVIRONMENTS button at the bottom of the window. In the ENVIRONMENT SETTINGS window, click on PROCESSING EXTENT. In the EXTENT section of the window that will appear select AS SPECIFIED BELOW from the drop down menu, and type in the following coordinates: TOP: 396000*, LEFT: -130000, RIGHT: 50000 and BOTTOM: 56000. Now click OK to close the ENVIRONMENTS SETTINGS window. Then click OK at the bottom of the POINT TO RASTER window.

Right click on the name of your newly created raster data layer (BND_RASTER) in the TABLE OF CONTENTS window and select PROPERTIES. **NOTE**: It may have been added below other data layers you already have in your data frame. Next, click on the SYMBOLOGY tab of the LAYER PROPERTIES window. In the left hand portion of the window, select UNIQUE VALUES. When it asks 'UNIQUE VALUES DO NOT EXIST. DO YOU WANT TO COMPUTE UNIQUE VALUES?' click YES. Next, click on the ADD ALL VALUES button. Double click on the coloured rectangle beside the number 1, and select black for the colour. Finally, click the OK button. All the cells in the BND_RASTER data layer where bottlenose dolphin were present will now be coloured black. All other cells will be classified as no data and will not be shown.

*The TOP value for the extent is slightly greater than that of the data frame (see exercise one). This ensures there is room for a whole number of grid cells between the bottom and the top of this extent.

Once you have completed this step, you can turn off both the BOTTLENOSE_DOLPHIN data layer and the BOTTLENOSE_DOLPHIN_ PROJECT data layer so that you can see your presence raster data layer properly. This is done by clicking on the small box with a tick in it next to their names in the TABLE OF CONTENTS window so that the tick disappears.

The contents of your MAP window should now look like this:

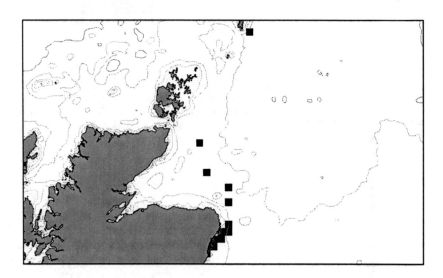

STEP 3: ADD THE SURVEY TRACK DATA LAYER TO THE GIS PROJECT:

In this step, a new data layer called SURVEY_TRACKS_NORTH_SEA.SHP will be added to your GIS project. This is a line data layer which gives information about the areas covered by the surveys from which the bottlenose dolphin locational data was collected. This will be used as the basis for determining which parts of the area were surveyed, but where bottlenose dolphin were not recorded. It is in a geographic projection and is based on the WGS 1984 datum.

This is based on the instruction sets called *How to add an existing data layer to a GIS project* and *How to change the display symbols for a data layer*. Generic versions of these instruction sets can be found in *An Introduction To Using GIS In Marine Biology*.

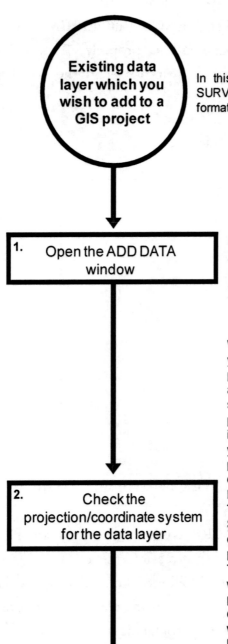

In this case the data layer you wish to add is called SURVEY_TRACKS_NORTH_SEA.SHP. It is already in a format which you can add to your GIS project.

Right click on the name of the data frame (LAYERS) in the TABLE OF CONTENTS window, and select ADD DATA, browse to the location of your data layer, select it and click ADD. For this exercise, the location is C:\GIS_EXERCISES\SURVEY_ TRACKS_NORTH _SEA.SHP.

Whenever you add a data layer to a GIS project, you should always check that is has a projection/coordinate system assigned to it, and look at what this projection/coordinate system is. This is so that you know whether you will need to assign a projection/coordinate system to it, or transform it into a different projection/coordinate system before you can use it in your GIS project. To check the projection/coordinate system of your newly added data layer, In the TOOLBOX window, go to DATA MANAGEMENT TOOLS> PROJECTIONS AND TRANSFORMATIONS> DEFINE PROJECTION. Select SURVEY_TRACKS_NORTH_SEA from the drop down menu in the top window. Check its projection/coordinate system in the lower window. This should be GCS_WGS_1984. **NOTE**: If you wish to find out more information about the specific projection/coordinate system of a data layer, you can click on the button at the end of the lower window to display its full details. Once you have looked at these details, click CANCEL to close the SPATIAL REFERENCES PROPERTIES window. Finally, click on the CANCEL button to close the DEFINE PROJECTION window.

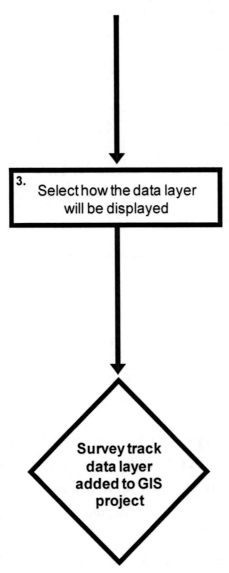

3. Select how the data layer will be displayed

Right click on the name of the SURVEY_TRACKS_NORTH_SEA in the TABLE OF CONTENTS window, and select PROPERTIES. In the LAYER PROPERTIES window, click on the SYMBOLOGY tab. On the left-hand side, under SHOW, select FEATURES> SINGLE SYMBOL. Next, under SYMBOL in the top middle of the LAYER PROPERTIES window, click on the button with the coloured line on it. This will open the SYMBOL SELECTOR window. On the right hand side of the SYMBOL SELECTOR window, click on the button next to COLOUR and hover the cursor over the fourth box down in the grey column. This will tell you that this is 30% grey. Select this by clicking on it. Next, select 3 for WIDTH. Finally, click OK to close the SYMBOL SELECTOR window and then OK to close the LAYER PROPERTIES window.

Survey track data layer added to GIS project

At the end of this step, the contents of your MAP window should look like the image at the top of the next page.

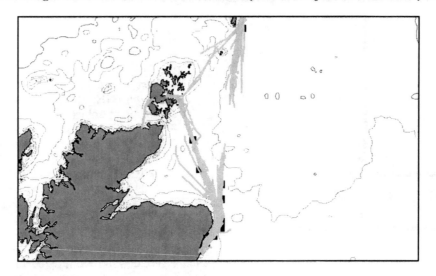

STEP 4: ADD A NEW FIELD TO THE ATTRIBUTE TABLE OF SURVEY_TRACKS_NORTH_SEA DATA LAYER AND FILL IT WITH THE NUMBER 0:

In order to make a presence-absence raster data layer, you first need to create a raster data layer based on the survey tracks. This will be done in step five. However, before you can do this you need have a field in the attribute table of your survey track data layer which has a value of zero in it for the legs of the survey track. This is created in step four.

The instruction set for this step is primarily based on one called *How to add a new field to an attribute table*, but it also includes steps from *How to use the field calculator tool to fill in values in a new field*. Generic versions of these instruction sets can be found in *An Introduction To Using GIS In Marine Biology*.

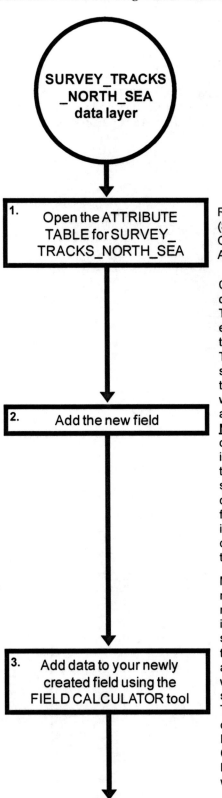

SURVEY_TRACKS _NORTH_SEA data layer

1. Open the ATTRIBUTE TABLE for SURVEY_ TRACKS_NORTH_SEA

2. Add the new field

3. Add data to your newly created field using the FIELD CALCULATOR tool

Right click on the name of the data layer (SURVEY_TRACKS_NORTH_SEA) in the TABLE OF CONTENTS window and select OPEN ATTRIBUTE TABLE.

Click the TABLE OPTIONS button at the top left corner of the TABLE window and select ADD FIELD. This opens the ADD FIELD window, where you will enter the required settings for your new field. Type the word SURVEY in the NAME window. In the TYPE window select SHORT INTEGER (since you simply want to add the number 0 to this field rather than use decimal places). In the PRECISION window, enter the number 16 (this allows you to put a maximum of 16 digits in this field). Now click OK. **NOTE**: Once a field has been created, you cannot change these properties or its name so it is important to get this right from the start. If you add the field and then you find that you have got the settings wrong, the best way to deal with this is to delete it and start again. **NOTE**: When you create a field, it will usually have values of zero for all lines in it. If this is the case, end this step here and move onto step five. If not, complete the instructions in this flow diagram.

Move the TABLE window so that you can see the main menu bar and optional toolbar areas of the main MAP window. Click on the CUSTOMIZE menu in the main menu bar and select TOOLBARS. Make sure there is a tick next to the EDITOR toolbar. In the EDITOR toolbar, click on the EDITOR button and select START EDITING. A new EDITOR window will then appear (usually on the right hand side of the main MAP window). Now click on the TABLE window and then right click on the field called SURVEY which you just created and select FIELD CALCULATOR This will open the FIELD CALCULATOR window. Type the number 0 into the lower window. Click OK. Finally, close the TABLE window.

4. Save your changes to the SURVEY_TRACKS_NORTH_SEA data layer

Once you have finished editing the attribute table for SURVEY_TRACKS_NORTH_SEA, you need to save the edits to your data layer. To do this, in the EDITOR toolbar, click on EDITOR and select SAVE EDITS. Next, in the EDITOR toolbar, click on the EDITOR button and select STOP EDITING.

New field added to attribute table of SURVEY_ TRACKS_NORTH _SEA with a value of 0 in it

When you have completed this step, the attribute table for the SURVEY_TRACKS_NORTH_SEA data layer should have changed from looking like the left half of the figure below to looking like the right half, with a new field called Survey added, which has a 0 in it for each line.

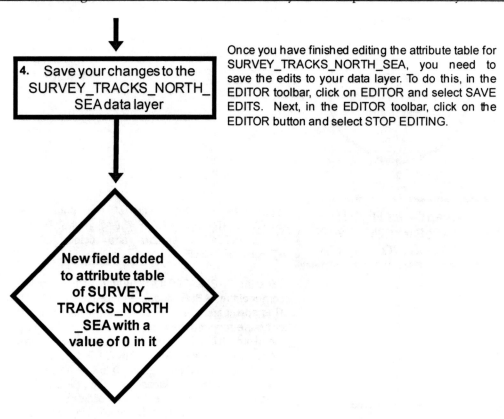

STEP 5: CONVERT SURVEY_TRACKS_NORTH_SEA DATA LAYER TO A RASTER DATA LAYER BASED ON THIS NEW FIELD:

In this step you will create a raster data layer where the cell values will indicate whether or not survey effort was conducted in a specific grid cell during the study. It will use the values in the SURVEY field created in step four to calculate the values for the cells, and when more than one survey leg falls within the same cell, it will use the value of the longest leg to determine the cell value. However, since all the survey legs in SURVEY_TRACKS_NORTH_SEA have a value of zero for the SURVEY field, this will result in a cell value of zero in all such cases regardless of the survey leg selected. These cells need to have a value of zero because you will want your final presence-absence data layer to have a zero value for cells which have been surveyed but where bottlenose dolphins were not recorded, and a value of one for those where bottlenose dolphin were recorded. In order to be able to do this, you must first create a raster data layer where all grid cells where there was any survey effort have a value of zero. This raster data layer (called TRACK_RASTER) will be created in this step.

The cell size, extent and projection/coordinate system of the raster data layer created in this step will all be the same as those used to create the data layer BND_RASTER in step two so that the two data layers overlay each other exactly. However, since the SURVEY_TRACKS_NORTH_SEA data layer is not in this projection/coordinate system, it will have to be transformed into it before the survey effort raster data layer is made.

These instructions are based on instruction sets called *How to transform data layers between different projections*, *How to create a new raster data layer from an existing line data layer* and *How to change the way a raster data layer is displayed*. Generic versions of these instruction sets can be found in *An Introduction To Using GIS In Marine Biology*.

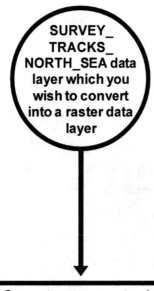

SURVEY_ TRACKS_ NORTH_SEA data layer which you wish to convert into a raster data layer

1. Convert your survey tracks data layer to the same projection/coordinate system as the data frame

In the TOOLBOX window, select DATA MANAGEMENT TOOLS> PROJECTIONS AND TRANSFORMATIONS> FEATURE> PROJECT. This will open the PROJECT window. Select SURVEY_TRACKS_NORTH_SEA from the drop down menu in the INPUT DATASET OR FEATURE CLASS window. In the OUTPUT DATASET OR FEATURE CLASS window enter C:\GIS_EXERCISES\SURVEY_ TRACKS_NORTH_SEA_PROJECT. Next, click on the button at the end of the OUTPUT COORDINATE SYSTEM window to open the SPATIAL REFERENCE PROPERTIES window. To select the custom projection/coordinate system being used for this exercise, click on the ADD COORDINATE SYSTEM button towards the top right hand corner of the SPATIAL REFERENCE PROPERTIES window and select IMPORT. This will open the BROWSE FOR DATASETS OR COORDINATE SYSTEMS window. Browse to the folder C:\GIS_EXERCISES\ and select BOTTLENOSE_DOLPHIN_PROJECT. Next click on the ADD button to close the BROWSE FOR DATASET window. In the SPATIAL REFERENCE PROPERTIES window, check that the projection/coordinate system which has been added is called NORTH SEA (look in the NAME section of the window). If so, click OK to close it. If not, click on the IMPORT button again and make sure you selected the right data layer to import the projection/coordinate system from. Finally, click OK to close the PROJECT window.

In the toolbox window, select CONVERSION TOOLS> TO RASTER> POLYLINE TO RASTER. In the POLYLINE TO RASTER window, select the line data layer called SURVEY_TRACKS_NORTH_SEA _PROJECT in the INPUT FEATURES window using the drop down menu. Select the field called SURVEY using the drop down menu in the VALUE FIELD window. Type C:\GIS_EXERCISES\TRACK_RASTER in the OUTPUT RASTER DATASET window. In the CELL ASSIGNMENT TYPE (OPTIONAL) window select MAXIMUM_LENGTH. Type the number 10000 into the CELLSIZE (OPTIONAL) window. Next, click on the ENVIRONMENTS button at the bottom of the window. In the ENVIRONMENT SETTINGS window, click on PROCESSING EXTENT. Select SAME AS LAYER BND_RASTER from the drop down menu that appears. Now click OK to close the ENVIRONMENTS SETTINGS window. Then click OK at the bottom of the POLYLINE TO RASTER window.

2. Convert your line data layer to a raster data layer using the POLYLINE TO RASTER tool

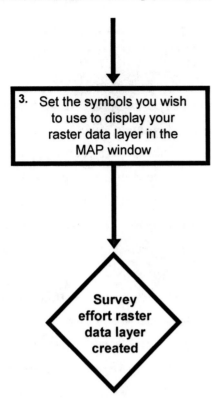

3. Set the symbols you wish to use to display your raster data layer in the MAP window

Right click on the name of your newly created raster data layer (TRACK_RASTER) in the TABLE OF CONTENTS window and select PROPERTIES. **NOTE**: It may have been added below other data layers you already have in your data frame. Next, click on the SYMBOLOGY tab of the LAYER PROPERTIES window. In the left hand portion of the SYMBOLOGY window, select UNIQUE VALUES. Next, click on the ADD ALL VALUES button. Double click on the coloured rectangle beside the number 0, and select black for the colour. Finally, click the OK button. All the cells in the TRACK_RASTER data layer where survey effort was conducted will now be coloured black.

Survey effort raster data layer created

Once you have completed this step, you can turn off all the data layers with the exception of LAND_NORTH_SEA, DEPTH_NORTH_SEA and TRACK_RASTER so that you can see the TRACK_RASTER data layer properly.

At the end of this step, the contents of your MAP window should look like this:

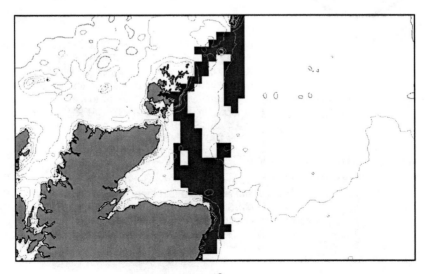

STEP 6: ADD THE TRACK_RASTER DATA LAYER AND BND_RASTER DATA LAYER TOGETHER:

The final step in creating a presence-absence raster data layer of species distribution is to add the raster data layers created in steps two and five together to give a raster data layer where the cell values are one for cells where bottlenose dolphins were recorded, and values of zero where survey effort was conducted, but no bottlenose dolphins were recorded. All other cells will be classified as No Data, and will not be displayed. This is done using the RASTER CALCULATOR tool. However, before it can be done, the presence raster data layer for bottlenose dolphins (BND_RASTER) needs to be reclassified so that all the No Data values in it have zero values. If this is not done, the process will not work properly. Therefore, this is the first thing which is done in the instruction set for this step. These instructions are based on instruction sets called *How to change cells to and from NO DATA in a raster data layer, How to create a new raster data layer by doing calculations with values of grid cells in another raster data layer* and *How to change the way a raster data layer is displayed*. Generic versions of these instruction sets can be found in *An Introduction To Using GIS In Marine Biology*.

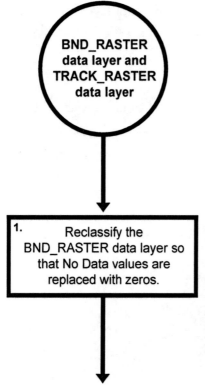

In the TOOLBOX window, select SPATIAL ANALYST TOOLS> RECLASS> RECLASSIFY. In the RECLASSIFY window, select BND_RASTER data layer from the drop down menu in the INPUT RASTER window. Select VALUE in the RECLASS FIELD window. In the RECLASSIFICATION section of the RECLASSIFY window, type 1 in the first line of the NEW VALUES column. Next type 0 into the second line of the NEW VALUES column (which currently has the words 'NO DATA' in it). This will change all the No Data values for the data layer to zeros. Scroll down and make sure that there is not a tick next to CHANGE MISSING VALUES TO NO DATA (OPTIONAL). Type C:\GIS_EXERCISES\BND_RASTER_2 in the OUTPUT RASTER window. Now click OK.

```
                                    In the TOOLBOX window, select SPATIAL ANALYST
                                    TOOLS> MAP ALGEBRA> RASTER CALCULATOR.
                                    This will open the RASTER CALCULATOR window. In
                                    the RASTER CALCULATOR window, you can build
                                    any expression you wish in the lower window. Existing
                                    raster data layers can be selected by double clicking
                                    on them in the LAYERS AND VARIABLES section of
                                    the window.  For this exercise, first double click on
   ┌─────────────────────────┐      BND_RASTER_2 to add it to the lower window. Then
   │ 2.    Open the RASTER    │      click the + sign.  Finally, double click on
   │  CALCULATOR tool and build│     TRACK_RASTER to add it to the expression.  The
   │  the expression which you │     final expression should read: "BND_RASTER_2" +
   │  want to use to create the new│  "TRACK_RASTER". NOTE: The expression has to be
   │      raster data layer    │     entered exactly like this, including the spaces before
   └─────────────────────────┘      and after the plus sign (+).  This expression will add
                                    the contents of the two data layers together.  This will
                                    produce a new raster data layer where there will be a
                                    value of 1 for cells where bottlenose dolphin were
                                    recorded and a zero for cells which were surveyed, but
                                    no bottlenose dolphins were recorded.  Finally, in the
                                    OUTPUT RASTER section of the window type
                                    C:\GIS_EXERCISES\PA_RASTER and click on the
                                    OK button to carry out this calculation.
```

In the TOOLBOX window, select SPATIAL ANALYST TOOLS> MAP ALGEBRA> RASTER CALCULATOR. This will open the RASTER CALCULATOR window. In the RASTER CALCULATOR window, you can build any expression you wish in the lower window. Existing raster data layers can be selected by double clicking on them in the LAYERS AND VARIABLES section of the window. For this exercise, first double click on BND_RASTER_2 to add it to the lower window. Then click the + sign. Finally, double click on TRACK_RASTER to add it to the expression. The final expression should read: "BND_RASTER_2" + "TRACK_RASTER". **NOTE**: The expression has to be entered exactly like this, including the spaces before and after the plus sign (+). This expression will add the contents of the two data layers together. This will produce a new raster data layer where there will be a value of 1 for cells where bottlenose dolphin were recorded and a zero for cells which were surveyed, but no bottlenose dolphins were recorded. Finally, in the OUTPUT RASTER section of the window type C:\GIS_EXERCISES\PA_RASTER and click on the OK button to carry out this calculation.

Right click on the name of your newly created raster data layer (PA_RASTER) in the TABLE OF CONTENTS window and select PROPERTIES. **NOTE**: It may have been added below other data layers you already have in your data frame. Next, click on the SYMBOLOGY tab of the LAYER PROPERTIES window. In the left hand portion of the SYMBOLOGY window, select UNIQUE VALUES. Next, click on the ADD ALL VALUES button. Double click on the coloured rectangle beside the number 1, and select black for the colour. Double click on the coloured rectangle beside the number 0 and select the second grey box down (which is 10% grey). Finally, click the OK button. All the cells in the PA_RASTER data layer where bottlenose dolphin were present will now be coloured black, while all surveyed cells where bottlenose dolphin were not recorded will be light grey. All unsurveyed cells will be left blank and so will not be visible.

3. Set the symbols you wish to use to display your raster data layer in the MAP window

Presence-Absence raster data layer created

At the end of this exercise, you should have the following data layers in your project:

1. **Bottlenose_Dolphin:** This is a point data layer containing locational records of bottlenose dolphins in the geographic projection. This data layer was made in exercise one.

2. **Bottlenose_Dolphin_Project:** This is a point data layer containing locational records of bottlenose dolphins converted into the custom transverse mercator projection of the data frame. This data layer was made in step two of this exercise.

3. **Survey_Tracks_North_Sea:** This is a line data layer containing the survey tracks in the geographic projection. This data layer was supplied for you to use and was added in step three of this exercise.

4. **Survey_Tracks_North_Sea_Project:** This is a line data layer containing the survey tracks converted into the custom transverse mercator projection of the data frame. This data layer was made in step five of this exercise.

5. **Depth_North_Sea:** This is a line data layer containing water depth contours. It is in the geographic projection. This data layer was supplied for you to use and was added as part of exercise one.

6. **Land_North_Sea:** This is a polygon data layer of the British Isles which indicates where the land is. It is in the geographic projection. This data layer was supplied for you to use and was added as part of exercise one.

7. **PA_Raster:** This is a raster data layer with a value of 1 for cells where bottlenose dolphins were recorded, and a value of 0 for cells which were surveyed but where bottlenose dolphin were not recorded. All other cells are classified as No Data and are not displayed. It is in the custom projection/coordinate system of the data frame. This data layer was made in step six of this exercise.

8. **BND_Raster:** This is a raster data layer with a value of 1 for cells where bottlenose dolphins were recorded. All other cells are classified as No Data and are not displayed. It is in the custom projection/coordinate system of the data frame. This data layer was made in step two of this exercise.

9. **BND_Raster_2:** This is a raster data layer with a value of 1 for cells where bottlenose dolphins were recorded, and a value of 0 for all other cells. It is in the custom projection/coordinate system of the data frame. This data layer was made in step six of this exercise.

10. Track_Raster: This is a raster data layer with a value of 0 for cells which were surveyed. All other cells are classified as No Data and are not displayed. It is in the custom projection/coordinate system of the data frame. This data layer was made in step five of this exercise.

If you are missing any of these data layers, return to the appropriate step for this exercise and make sure you complete it properly. Finally, make sure that only Depth_North_Sea, Land_North_Sea, and PA_Raster are set to display (i.e. have a tick in the small box next to their names). For all other data layers, click on this box to remove the tick so that they are not displayed.

At the end of this exercise, the contents of your MAP window should look like this.

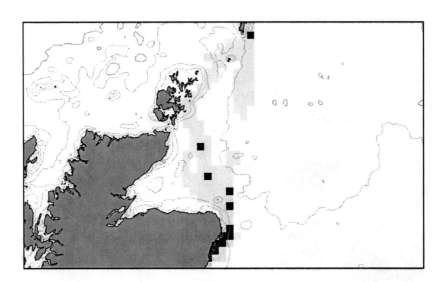

Optional Extra:

If you wish to get more experience in creating presence-absence raster data layers, you can use the data layer HARBOUR_PORPOISE.SHP, which contains locations where harbour porpoises were recorded in the same surveys, to create a presence-absence grid for harbour porpoises using the same instructions. However, when you do this you will notice that you do not need to create the TRACK_RASTER data layer again as you can use the one you produced while creating the bottlenose dolphin presence-absence raster data layer. In addition, you will need to remember to call your final presence-absence raster data layer by another name so that you know what is in it and so that it will not clash with the one you created for bottlenose dolphin. When you have finished creating the presence-absence raster data layer for harbour porpoises, and if you have done everything correctly, the contents of your MAP window should look like the figure provided below.

NOTE: Remember you may need to turn some data layers off once you have created your harbour porpoise presence-absence raster to be able to see it properly and to make it look exactly like this.

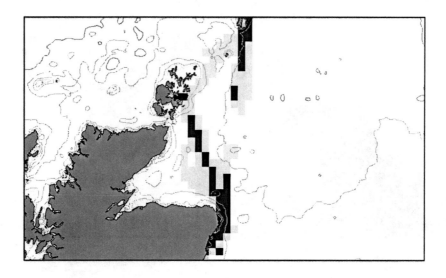

In terms of using presence-absence raster data layers to help answer research questions, you might be interested in looking at how the distributions of species differ within a study area. By comparing the presence-absence raster data layers you have created for bottlenose dolphin and for harbour porpoises, you can see that these two species clearly have different distributions in northeast Scotland. Harbour porpoises occupy a greater number of grid cells and, in particular, occur in more offshore grid cells than bottlenose dolphins, which are primarily restricted to coastal grid cells towards the southern end of the study area.

Chapter Five

Exercise Three: Creating A Species Richness Raster Data Layer From Survey Data

In exercise two, you produced a presence-absence raster data layer which showed where bottlenose dolphin were recorded in relation to those areas which were surveyed. This is fine for looking at the distribution of individual species. However, you may be interested not only in the distribution of individual species, but in identifying the areas with the highest diversity of species. Therefore, in this exercise, you will create a species richness raster data layer for all species of cetacean recorded in the surveys from northeast Scotland. The values for cells in this raster data layer will indicate whether the grid cell was surveyed and how many different species were recorded in it during the study.

For this exercise, you will use three data layers. Two of these were created in exercise two, so you will need to have successfully completed this exercise first before trying to do this one. The third can be found in the data download for this book which is available from www.gisinecology.com/books/marinebiologysupplementaryworkbook. These data layers should all be saved in a folder on your C drive called GIS_EXERCISES prior to starting this exercise, so that it has the address C:\GIS_EXERCISES\.

The data layers needed for this exercise are:

1. PA_Raster: This is a raster data layer of bottlenose dolphin presence-absence created in exercise two. In this data layer, grid cells where bottlenose dolphins were recorded have a value of one, grid cells which were surveyed but where bottlenose dolphins were not recorded have a value of zero. Unsurveyed cells are classified as No Data. This data layer is in the custom transverse mercator projection used for the data frame in other exercises and is based on the WGS 1984 datum.

2. Track_Raster: This is a raster data layer of survey effort created in exercise two. Grid cells which were surveyed have a value of zero, while all unsurveyed grid cells are classified as No Data and so are not displayed. This data layer is in the custom transverse mercator projection used for the data frame in other exercises and is based on the WGS 1984 datum.

3. All_Species.shp: This is the point data layer of locational records for all species recorded during the surveys. Each point represents a record for a single group of cetaceans of a single species recorded during the surveys. It is in the geographic projection and is based on the WGS 1984 datum.

The starting point for this exercise is the GIS project created in exercise two. Open the ArcMap module of the ArcGIS 10.1 software. When it opens, you will be presented with a window which has the heading ARCMAP – GETTING STARTED. To open an existing GIS project, click on EXISTING MAPS in the directory tree on the left hand side and then select the project called EXERCISE_TWO in the right hand section of the window. Now, click OPEN at the bottom of this window. If it is not listed here, click on the BROWSE FOR MORE option on the left had side of the window and use the browser window to find and select this GIS project from the folder C:\GIS_EXERCISES\. Now, click OPEN at the bottom of this window. If the ARCMAP – GETTING STARTED window does not appear when you start the ArcMap module, you can open an existing project by clicking on FILE from the main menu bar area, and selecting OPEN. When the OPEN window opens, browse to the location where your project is saved (C:\GIS_EXERCISES\), select it and click OPEN.

Once you have opened the GIS project called EXERCISE_TWO, the first thing you need to do is save it under a new name. This is because you do not want to alter the contents of the original project, you just want to base your new one on it since this saves you having to add all the data layers again, and also having to reset the projection/coordinate system of the data frame. To save the project under a new name, click on FILE from the main menu bar area, and select SAVE AS. Save your GIS project as EXERCISE_THREE in the C:\GIS_EXERCISES\folder.

The data layers which you need in your GIS project to start exercise three are: 1. Depth_North_Sea.shp; 2. Land_North_Sea; 3. PA_Raster; 4. Track_Raster. The land and depth data layers will not be used in this exercise, but are useful as they show other features in the study area. You will need to remove any other data layers from your GIS project. This is done by right-clicking on the name of the data layer you wish to remove in the TABLE OF CONTENTS window, and selecting REMOVE. **NOTE**: This will not delete these data layers from the folder C:\GIS_EXERCISES\, it will only remove them from your GIS project. However, you need to keep the source files for these data layers in the C:\GIS_EXERCISES\ as you may need to use them for other exercises later in the book.

Thus, at the start of this exercise, the contents of your TABLE OF CONTENTS window should look like this:

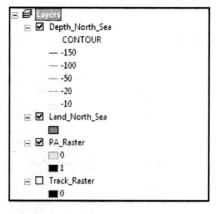

And you MAP window should look like this:

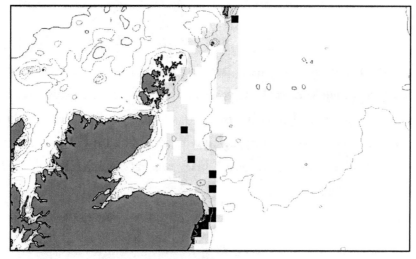

Summary Flow Diagram For Creating A Species Richness Raster Data layer

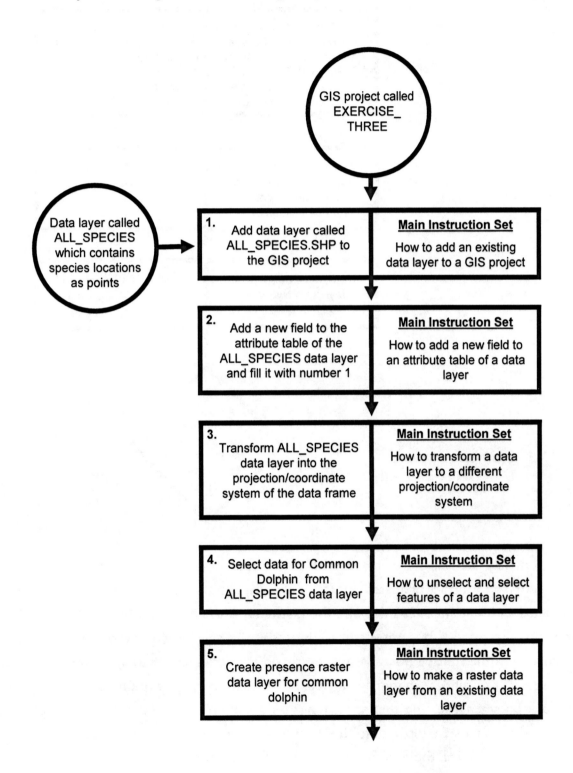

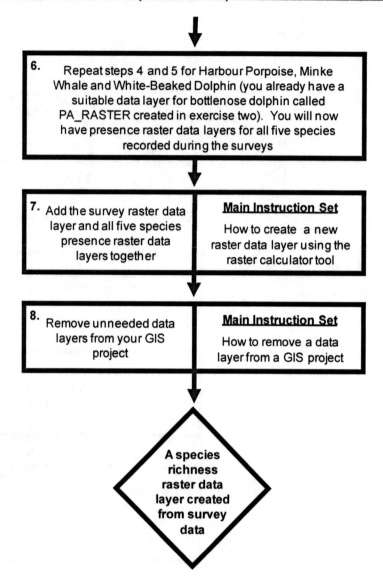

Once you have familiarised yourself with this summary flow diagram, read through the instruction set for the first step in the flow diagram in its entirety. When you have completed the first step, read through the instructions again and make sure that you have completed it properly. It is important to do this at this stage as you need to use the results of one step as the starting point for the next, and it is much easier to spot where you have gone wrong at the end of an individual step, rather than trying to work it out later when you get stuck or when you find that something has gone wrong. Once you have successfully completed the first step, move onto the second step and repeat this process, and so on until you have completed all the steps in the summary flow diagram.

At various points, images of the contents of the MAP window, the LAYOUT window, the TABLE OF CONTENTS window and/or the TABLE window will be provided so that you have an idea of what your GIS project should look like at that specific point as you progress through this exercise.

Instruction Sets For The Individual Steps Identified In The Summary Flow Diagram:

STEP 1: ADD DATA LAYER CALLED ALL_SPECIES.SHP TO THE GIS PROJECT:

In this step, a new data layer called ALL_SPECIES.SHP will be added to the project. This is a point data layer which gives information about the locations where all five species recorded in this study are seen. This will be used as the basis for creating individual presence raster data layers for the species other than bottlenose dolphin (since you already created a presence-absence data layer , called PA_RASTER, for this species in exercise two which can be used for this exercise). ALL_SPECIES.SHP is in a geographic projection and is based on the WGS 1984 datum. These instructions are based on instruction sets called *How to add an existing data layer to a GIS project* and *How to change the display symbols for a data layer* from *An Introduction To Using GIS In Marine Biology*.

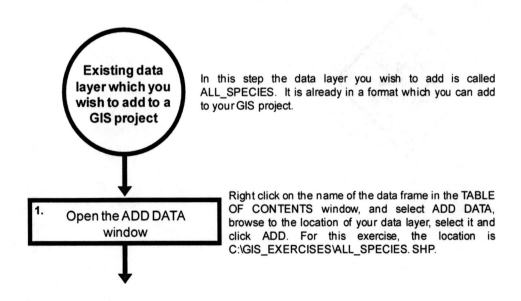

Existing data layer which you wish to add to a GIS project

In this step the data layer you wish to add is called ALL_SPECIES. It is already in a format which you can add to your GIS project.

1. Open the ADD DATA window

Right click on the name of the data frame in the TABLE OF CONTENTS window, and select ADD DATA, browse to the location of your data layer, select it and click ADD. For this exercise, the location is C:\GIS_EXERCISES\ALL_SPECIES.SHP.

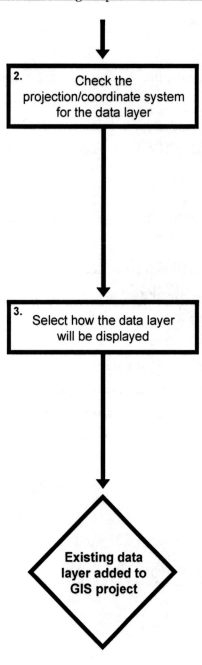

2. Check the projection/coordinate system for the data layer

3. Select how the data layer will be displayed

Existing data layer added to GIS project

The first thing you need to do is to check that this data layer has the correct projection/coordinate system assigned to it. In the TOOLBOX window, go to DATA MANAGEMENT TOOLS> PROJECTIONS AND TRANSFORMATIONS> DEFINE PROJECTION. Select ALL_SPECIES from the drop down menu in the top window. Check its projection/coordinate system in the lower window. This should be GCS_WGS_1984. Now click CANCEL.

Right click on the name of the ALL_SPECIES data layer in the TABLE OF CONTENTS window, and select PROPERTIES. In the LAYER PROPERTIES window, click on the SYMBOLOGY tab. On the left-hand side, under SHOW, select CATEGORIES> UNIQUE VALUES. Next, under VALUE FIELD select SPECIES from the drop down menu. Then click on the ADD ALL VALUES button. This will add all five species names to the lower window. You can now double click on the symbol next to each species name and change how it looks in the SYMBOL SELECTOR window which will open. When doing this, select CIRCLE 2 for the symbol design, and select 12 for the size. Next, select the appropriate colour for the species (see below). When you have selected the symbol, size and colour for a species, click OK to close the SYMBOL SELECTOR window and repeat this process for each species in turn. In terms of colours, select red for the colour of the Bottlenose Dolphin symbol, blue for Harbour Porpoises, green for Minke Whales, yellow for Common Dolphin and white for the White-Beaked Dolphin. Once all five species have been done, click OK to close the LAYER PROPERTIES window.

Once you have completed this step, the contents of your MAP window should look like this:

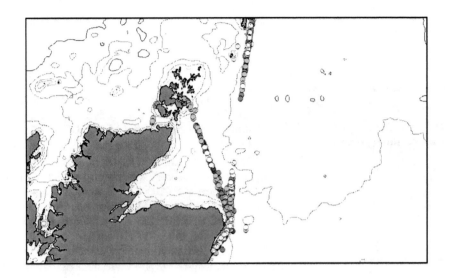

STEP 2: ADD A NEW FIELD TO THE ATTRIBUTE TABLE OF THE ALL_SPECIES DATA LAYER AND FILL IT WITH THE NUMBER ONE:

In order to make a species richness raster data layer, you first need to create a presence raster data layer for each species, this will be done in steps five and six. However, before you can do this you need to have a field in the attribute table of your species location data layer which has a value of one in it for all locational records. This indicates species presence and is created in this step. The instructions for this step are primarily based on one called *How to add a new field to an attribute table*, but it also includes steps from *How to use the field calculator tool to fill in values in a new field*. The generic versions of these instruction sets can be found in *An Introduction To Using GIS In Marine Biology*.

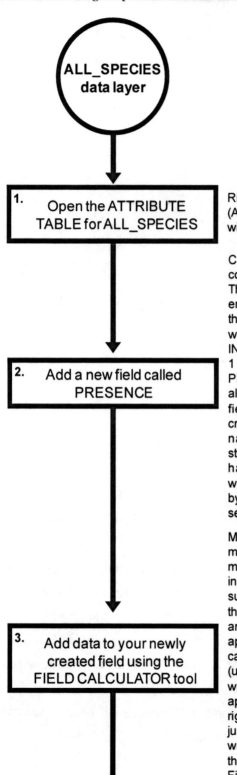

1. Open the ATTRIBUTE TABLE for ALL_SPECIES

Right click on the name of the data layer (ALL_SPECIES) in the TABLE OF CONTENTS window and select OPEN ATTRIBUTE TABLE.

Click the TABLE OPTIONS button at the top left corner of the TABLE window and select ADD FIELD. This opens the ADD FIELD window where you will enter the required settings for your new field. Type the word PRESENCE in the NAME section of the window. In the TYPE section select SHORT INTEGER (since you simply want to add the number 1 to this field rather than use decimal places). In the PRECISION section, enter the number 16 (this allows you to put a maximum of sixteen digits in this field. Now click OK. **NOTE**: Once a field has been created, you cannot change these properties or its name so it is important to get this right from the start. If you add the field and then you find that you have got the settings wrong, the best way to deal with this is to delete it and start again. This is done by right clicking on the name of the field and selecting DELETE FIELD.

2. Add a new field called PRESENCE

Move the TABLE window so that you can see the main menu bar and optional toolbar areas of the main MAP window. Click on the CUSTOMIZE menu in the main menu bar and select TOOLBARS. Make sure there is a tick next to the EDITOR toolbar. In the EDITOR toolbar, click on the EDITOR button and select START EDITING. If a warning window appears, this is okay. Just Click CONTINUE and carry on. A new EDITOR window may then appear (usually on the right hand side of the main MAP window – do not worry if this window does not appear). Now click on the TABLE window and then right click on the field called PRESENCE which you just created and select FIELD CALCULATOR This will open the FIELD CALCULATOR window. Type the number 1 into the lower window. Click OK. Finally, close the TABLE window.

3. Add data to your newly created field using the FIELD CALCULATOR tool

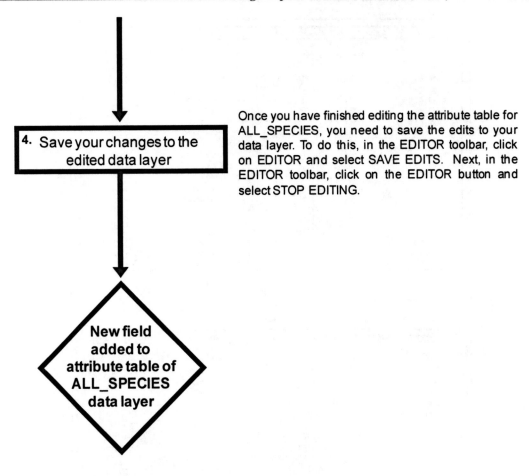

4. Save your changes to the edited data layer

Once you have finished editing the attribute table for ALL_SPECIES, you need to save the edits to your data layer. To do this, in the EDITOR toolbar, click on EDITOR and select SAVE EDITS. Next, in the EDITOR toolbar, click on the EDITOR button and select STOP EDITING.

New field added to attribute table of ALL_SPECIES data layer

When you have completed this step, the attribute table for the ALL_SPECIES data layer should have changed from looking like the top half of the figure on the next page to looking like the bottom half, with a new field called PRESENCE added, which has a 1 in it for each line. If the attribute table is not already open, right click on the name ALL_SPECIES in the TABLE OF CONTENTS window and select OPEN ATTRIBUTE TABLE to check that it now looks like the lower figure.

FID	Shape *	Sighting_I	Latitude	Longitude	Species	Number
0	Point	500	57.148333	-2.069667	Bottlenose dolphin	15
1	Point	501	57.1515	-2.045167	Bottlenose dolphin	5
2	Point	502	57.1525	-2.042667	Bottlenose dolphin	5
3	Point	506	57.141667	-2.070833	Bottlenose dolphin	6
4	Point	511	57.148	-2.051667	Bottlenose dolphin	6
5	Point	535	57.151167	-2.045333	Bottlenose dolphin	4
6	Point	554	57.149667	-2.05	Bottlenose dolphin	9
7	Point	591	58.472667	-2.27	Bottlenose dolphin	2
8	Point	592	57.15	-2.034667	Bottlenose dolphin	6
9	Point	595	57.157167	-2.033333	Bottlenose dolphin	6
10	Point	596	57.150833	-2.046833	Bottlenose dolphin	10
11	Point	600	57.146833	-2.054833	Bottlenose dolphin	7
12	Point	601	57.262333	-1.897333	Bottlenose dolphin	1
13	Point	603	57.373167	-1.7365	Bottlenose dolphin	1
14	Point	604	57.783333	-1.636333	Bottlenose dolphin	2
15	Point	632	57.147167	-2.059	Bottlenose dolphin	4
16	Point	657	57.455667	-1.6355	Bottlenose dolphin	2
17	Point	658	57.381333	-1.719667	Bottlenose dolphin	1
18	Point	659	57.381333	-1.719667	Bottlenose dolphin	1
19	Point	711	57.285333	-1.8405	Bottlenose dolphin	6
20	Point	717	57.143	-2.065333	Bottlenose dolphin	15

FID	Shape *	Sighting_I	Latitude	Longitude	Species	Number	Presence
0	Point	500	57.148333	-2.069667	Bottlenose dolphin	15	1
1	Point	501	57.1515	-2.045167	Bottlenose dolphin	5	1
2	Point	502	57.1525	-2.042667	Bottlenose dolphin	5	1
3	Point	506	57.141667	-2.070833	Bottlenose dolphin	6	1
4	Point	511	57.148	-2.051667	Bottlenose dolphin	6	1
5	Point	535	57.151167	-2.045333	Bottlenose dolphin	4	1
6	Point	554	57.149667	-2.05	Bottlenose dolphin	9	1
7	Point	591	58.472667	-2.27	Bottlenose dolphin	2	1
8	Point	592	57.15	-2.034667	Bottlenose dolphin	6	1
9	Point	595	57.157167	-2.033333	Bottlenose dolphin	6	1
10	Point	596	57.150833	-2.046833	Bottlenose dolphin	10	1
11	Point	600	57.146833	-2.054833	Bottlenose dolphin	7	1
12	Point	601	57.262333	-1.897333	Bottlenose dolphin	1	1
13	Point	603	57.373167	-1.7365	Bottlenose dolphin	1	1
14	Point	604	57.783333	-1.636333	Bottlenose dolphin	2	1
15	Point	632	57.147167	-2.059	Bottlenose dolphin	4	1
16	Point	657	57.455667	-1.6355	Bottlenose dolphin	2	1
17	Point	658	57.381333	-1.719667	Bottlenose dolphin	1	1
18	Point	659	57.381333	-1.719667	Bottlenose dolphin	1	1
19	Point	711	57.285333	-1.8405	Bottlenose dolphin	6	1
20	Point	717	57.143	-2.065333	Bottlenose dolphin	15	1

Now close the attribute table.

STEP 3: TRANSFORM ALL_SPECIES DATA LAYER INTO THE PROJECTION/COORDINATE SYSTEM OF THE DATA FRAME:

When you come to creating presence raster data layers for the remaining species, these will need to be created in the same projection/coordinate system as the data frame so that they will match up with the TRACK_RASTER and the presence-absence raster for bottlenose dolphins (PA_RASTER) created in exercise two. However, since the ALL_SPECIES data layer is not in this projection/coordinate system, you will need to transform it into the same projection/coordinate system of the data frame before you can start making any presence raster data layers from it. In exercise two, this was done as part of step two when the presence raster data layer for bottlenose dolphin was created. However, since you are going to create multiple presence rasters from the same point data layer this time round, you need to transform the ALL_SPECIES data layer to the right projection in a separate step.

The instructions for this step are based on instruction sets called *How to transform data layers between different projections* and *How to change the display symbols for a data layer* from *An Introduction To Using GIS In Marine Biology*.

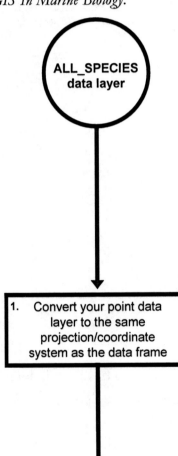

In the TOOLBOX window, select DATA MANAGEMENT TOOLS> PROJECTIONS AND TRANSFORMATIONS> FEATURE> PROJECT. This will open the PROJECT window. Select ALL_SPECIES from the drop down menu in the INPUT DATASET OR FEATURE CLASS window. In the OUTPUT DATASET OR FEATURE CLASS window enter C:\GIS_EXERCISES\ALL_SPECIES_PROJECT. Since you created the custom projection/coordinate system which you wish to transform the ALL_SPECIES data layer into in exercise two, you can choose to import it rather than having to make it from scratch again.

To select the custom projection/coordinate system being used for this exercise, click on the button at the right hand end of the OUTPUT COORDINATE SYSTEM section. This will open the SPATIAL REFERENCE PROPERTIES window. Next, click on the ADD COORDINATE SYSTEM button towards the top right hand corner of the SPATIAL REFERENCE PROPERTIES window and select IMPORT. This will open the BROWSE FOR DATASETS OR COORDINATE SYSTEMS window. Browse to the folder C:\GIS_EXERCISES\ and select BOTTLENOSE_DOLPHIN_PROJECT. Next click on the ADD button to close the BROWSE FOR DATASET window. In the SPATIAL REFERENCE PROPERTIES window, check that the projection/coordinate system which has been added is called NORTH SEA (look in the NAME section of the window). If so, click OK to close it. If not, click on the IMPORT button again and make sure you selected the right data layer to import the projection/coordinate system from. Finally, click OK to close the PROJECT window.

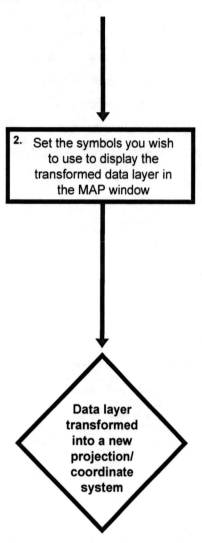

2. Set the symbols you wish to use to display the transformed data layer in the MAP window

Right click on the name of the transformed point data layer (ALL_SPECIES_PROJECT) in the TABLE OF CONTENTS window and select PROPERTIES. Next, click on the SYMBOLOGY tab of the LAYER PROPERTIES window. In the left hand portion of the LAYER PROPERTIES window, select CATEGORIES> UNIQUE VALUES. You will want to use the same symbols as you used for the ALL_SPECIES data layer. These can be loaded in by clicking on the IMPORT button on the right hand side of the LAYER PROPERTIES window. In the IMPORT SYMBOLOGY window which will open as a result, select ALL_SPECIES from the drop down menu and click on the OK button. In the IMPORT SYMBOLOGY MATCHING DIALOG window which will open, select SPECIES as the VALUE FIELD, and then click OK. Finally, click OK to close the LAYER PROPERTIES window.

Data layer transformed into a new projection/ coordinate system

To check that you have done this step properly, right click on the name of your new data layer (ALL_SPECIES_PROJECT) in the TABLE OF CONTENTS window and select properties. Click on the SOURCE tab and make sure that the contents of the DATA SOURCE section of the window has the following text in it (you may have to scroll down to see it all):

Projected Coordinate System: NORTH SEA
Projection: Transverse_Mercator
False_Easting: 0.00000000
False_Northing: 0.00000000

Central_Meridian: -1.00000000
Scale_Factor: 1.00000000
Latitude_Of_Origin: 56.50000000
Linear Unit: Meter

Geographic Coordinate System: GCS_WGS_1984
Datum: D_WGS_1984
Prime Meridian: Greenwich
Angular Unit: Degree

If it does not, you will need to repeat this step to ensure that you have assigned the correct projection/coordinate system to your data frame.

STEP 4: SELECT DATA FOR COMMON DOLPHIN FROM ALL_SPECIES DATA LAYER:

In exercise two, you converted the entire contents of a data layer into a raster data layer to create a presence raster data layer for bottlenose dolphins. However, the point data layer which you did this from only contained locations of bottlenose dolphin sightings. In this exercise, we wish to create presence raster data layers for individual species from a point data layer which has information about multiple species in it (ALL_SPECIES_PROJECT). In order to ensure that only information about a single species is converted into each presence raster data layer, you need to select the data for that species in some way. This is done using the SELECT BY ATTRIBUTE tool. In this step, you will initially select just the data for common dolphin.

This is done using the following instructions, which are based on ones called *How to unselect and select features of a data layer* and *How to change the display symbols for a data layer* from *An Introduction To Using GIS In Marine Biology*.

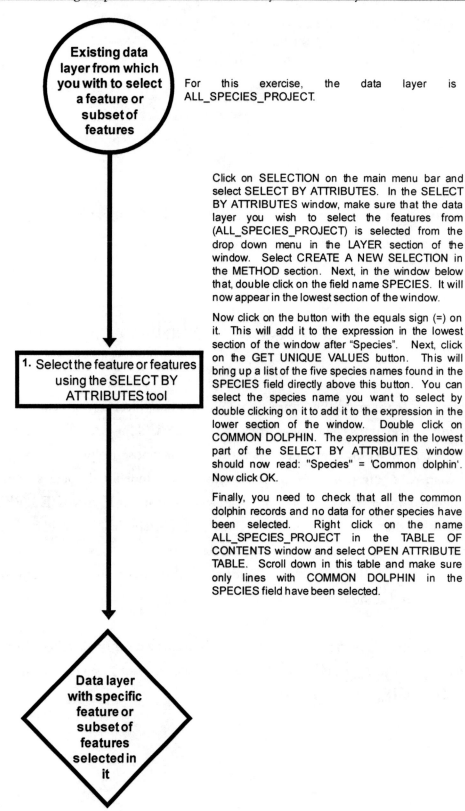

Existing data layer from which you with to select a feature or subset of features

For this exercise, the data layer is ALL_SPECIES_PROJECT.

Click on SELECTION on the main menu bar and select SELECT BY ATTRIBUTES. In the SELECT BY ATTRIBUTES window, make sure that the data layer you wish to select the features from (ALL_SPECIES_PROJECT) is selected from the drop down menu in the LAYER section of the window. Select CREATE A NEW SELECTION in the METHOD section. Next, in the window below that, double click on the field name SPECIES. It will now appear in the lowest section of the window.

Now click on the button with the equals sign (=) on it. This will add it to the expression in the lowest section of the window after "Species". Next, click on the GET UNIQUE VALUES button. This will bring up a list of the five species names found in the SPECIES field directly above this button. You can select the species name you want to select by double clicking on it to add it to the expression in the lower section of the window. Double click on COMMON DOLPHIN. The expression in the lowest part of the SELECT BY ATTRIBUTES window should now read: "Species" = 'Common dolphin'. Now click OK.

1. Select the feature or features using the SELECT BY ATTRIBUTES tool

Finally, you need to check that all the common dolphin records and no data for other species have been selected. Right click on the name ALL_SPECIES_PROJECT in the TABLE OF CONTENTS window and select OPEN ATTRIBUTE TABLE. Scroll down in this table and make sure only lines with COMMON DOLPHIN in the SPECIES field have been selected.

Data layer with specific feature or subset of features selected in it

At the end of this step, the attribute table for ALL_SPECIES_PROJECT should look like this (with the common dolphin data selected and the data for all other species not selected – you may need to scroll down to see which records are selected and which are not):

FID	Shape *	Sighting_I	Latitude	Longitude	Species	Number	Presence
70	Point	732	57.144	-2.062	Bottlenose dolphin	2	1
71	Point	750	57.149667	-2.048167	Bottlenose dolphin	5	1
72	Point	751	57.149167	-2.049667	Bottlenose dolphin	10	1
73	Point	752	57.1465	-2.056167	Bottlenose dolphin	8	1
74	Point	757	57.144833	-2.061333	Bottlenose dolphin	1	1
75	Point	758	57.145333	-2.058667	Bottlenose dolphin	5	1
76	Point	809	57.443667	-1.6705	Bottlenose dolphin	4	1
77	Point	876	58.1335	-2.021867	Bottlenose dolphin	2	1
78	Point	877	58.157167	-2.0425	Bottlenose dolphin	3	1
79	Point	878	58.159167	-2.044167	Bottlenose dolphin	2	1
89	Point	27	57.333167	-1.785333	Harbour porpoise	1	1

Now close the attribute table.

The contents of your MAP window should look like this:

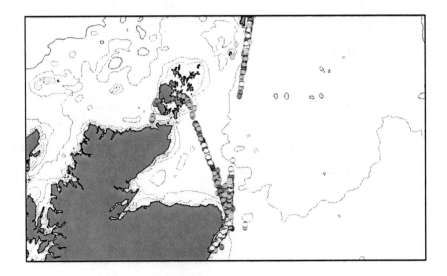

STEP 5: CREATE PRESENCE RASTER DATA LAYER FOR COMMON DOLPHIN:

In this step you will create a raster data layer where the cell values will indicate whether common dolphin were recorded as present during the study. It will use the values in the PRESENCE field of ALL_SPECIES_PROJECT. However, it will only base the new raster data layer on the selected features (i.e. those of common dolphin selected in step four). When more than one common dolphin record falls within the same cell, it will calculate an average value for that cell. Since all the common dolphin records in ALL_SPECIES_PROJECT data layer have a value of one for the presence field, this will result in a cell value of one in all such cases. The same cell size, extent and projection/coordinate system will be used for this raster data layer as for the other raster data layers created so far in these exercises. This will ensure that they all directly overlay each other so that they can be added together to create a species richness raster data layer in step seven.

Two raster data layers will be created in this step. In the first, which for common dolphin will be called CD_RASTER, cells where common dolphin were recorded will have a value of one, and all other cells will be classified as NO DATA, and will not be displayed. The second, which will be based on this first raster data layer, and that for common dolphin, will be called CD_RASTER_2, and will have cell values of one for cells where common dolphin were recorded, and values of zero, rather than NO DATA, for all other cells.

These instructions are based on instruction sets called *How to change cells to and from NO DATA in a raster data layer*, *How to create a new raster data layer from an existing point data layer* and *How to change the way a raster data layer is displayed*. The generic versions of these instruction sets can be found in *An Introduction To Using GIS In Marine Biology*.

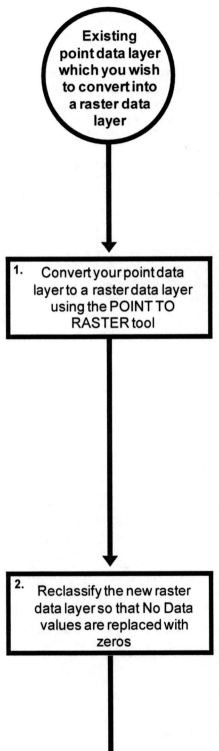

In the toolbox window, select CONVERSION TOOLS > TO RASTER > POINT TO RASTER. In the POINT TO RASTER window, select the point data layer called ALL_SPECIES_PROJECT in the INPUT FEATURES window using the drop down menu. Select the field called PRESENCE using the drop down menu in the VALUE FIELD window. Type C:\GIS_EXERCISES\ CD_RASTER in the OUTPUT RASTER DATASET window. In the CELL ASSIGNMENT TYPE (OPTIONAL) window select MEAN. Type the number 10000 into the CELLSIZE (OPTIONAL) window. Next, click on the ENVIRONMENTS button at the bottom of the window. In the ENVIRONMENT SETTINGS window, click on PROCESSING EXTENT. In the EXTENT section of the window that will appear select SAME AS LAYER TRACK_RASTER. This will match the extent of the new raster data layer to the extent of this existing one. Click OK to close the ENVIRONMENTAL SETTINGS window. Finally, click OK at the bottom of the POINT TO RASTER window.

Right-click on the name of your raster data layer (CD_RASTER) in the TABLE OF CONTENTS window and select PROPERTIES. In the LAYER PROPERTIES window, click on the SYMBOLOGY TAB, select UNIQUE VALUES and then click OK to close this window. Next, in the TOOLBOX window, select SPATIAL ANALYST TOOLS> RECLASS> RECLASSIFY. In the RECLASSIFY window, select CD_RASTER data layer from the drop down menu in the INPUT RASTER window. Select VALUE in the RECLASS FIELD window. In the RECLASSIFICATION section of the RECLASSIFY window, type 1 in the first line of the NEW VALUES column. Next type 0 into the second line of the NEW VALUES column (which currently has the words 'NO DATA' in it). This will change all the No Data values for the data layer to zeros. Scroll down and make sure that there is no tick next to CHANGE MISSING VALUES TO NO DATA (OPTIONAL). Type C:\GIS_EXERCISES\CD_ RASTER_2 in the OUTPUT RASTER window. Now click OK.

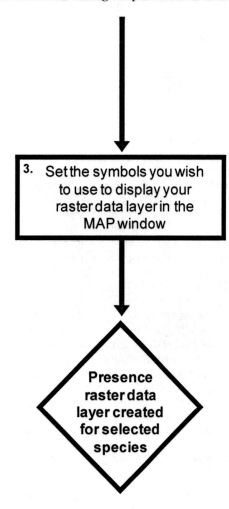

3. Set the symbols you wish to use to display your raster data layer in the MAP window

Presence raster data layer created for selected species

Right click on the name of your newly created raster data layer (CD_RASTER_2) in the TABLE OF CONTENTS window and select PROPERTIES. **NOTE**: It may have been added below other data layers you already have in your data frame. Next, click on the SYMBOLOGY tab of the LAYER PROPERTIES window. In the left hand portion of the SYMBOLOGY window, select UNIQUE VALUES. Next, click on the ADD ALL VALUES button. Double click on the coloured rectangle beside the number 1, and select black for the colour. Do the same for the coloured rectangle next to zero, but select 10% grey. Finally, click the OK button. All the cells in the CD_RASTER_2 data layer where common dolphin were present will now be coloured black, all other cells will be light grey.

At the end of this step, make sure that only the LAND_NORTH_SEA, DEPTH_NORTH_SEA and CD_RASTER_2 data layers are set to display (i.e. have ticks next to their names in the TABLE OF CONTENTS window). The contents of your MAP window should then look like the image at the top of the next page.

74

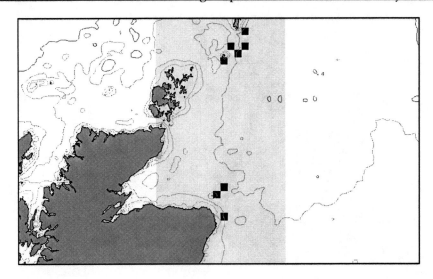

:

STEP 6: REPEAT STEPS 4 AND 5 FOR ALL OTHER SPECIES TO CREATE A SEPARATE PRESENCE DATA LAYER FOR EACH SPECIES:

In order to create presence rasters data layers for all other species, steps four and five are repeated for the harbour porpoise (creating raster data layers called HP_RASTER and HP_RASTER_2 in step five), minke whale (creating raster data layers called MW_RASTER and MW_RASTER_2), and white-beaked dolphin (creating raster data layers called WBD_RASTER and WBD_RASTER_2). You do not need to create a new presence raster data layer for bottlenose dolphin as you already created one in exercise two, and this is already loaded into your GIS project as PA_RASTER.

The raster data layer HP_RASTER_2 created in this step, should look like the image at the top of the next page.

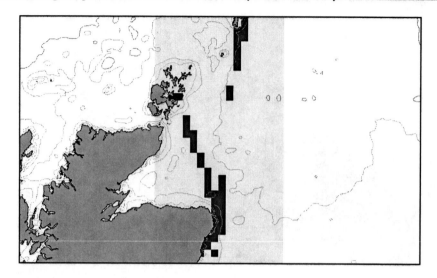

The raster data layer MW_RASTER_2 created in this step, should look like the image below:

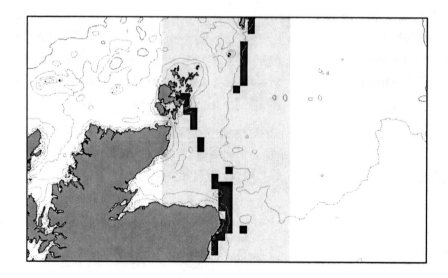

The raster data layer WBD_RASTER_2 created in this step, should look like this:

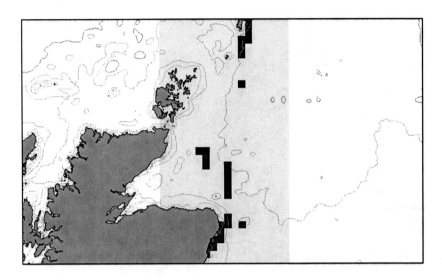

STEP 7: ADD THE SURVEY RASTER DATA LAYER AND ALL FIVE SPECIES PRESENCE RASTER DATA LAYERS TOGETHER:

The next step in creating a species richness raster data layer is to add together all the individual presence raster data layers for each species (PA_RASTER – which is the bottlenose dolphin presence-absence raster layer created in exercise two, CD_RASTER_2 – which was created in steps five, the data layers HP_RASTER_2, MW_RASTER_2 and WBD_RASTER_2 – which were created in step six), and the survey track raster data layer (TRACK_RASTER). At the end of this step, you will have a raster data layer where the cell values range from zero for grid cells where no species were recorded but where there was survey effort, to five where all five species were recorded. This step is achieved using the raster calculator tool.

These instructions are based on instruction sets called *How to create a new raster data layer by doing calculations with values of grid cells in another raster data layer* and *How to change the way a raster data layer is displayed*. Generic versions of these instruction sets can be found in *An Introduction To Using GIS In Marine Biology*.

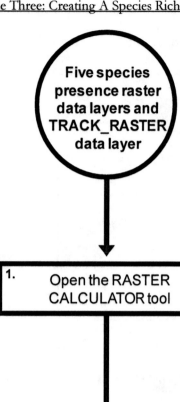

Five species presence raster data layers and TRACK_RASTER data layer

1. Open the RASTER CALCULATOR tool

In the TOOLBOX window, select SPATIAL ANALYST TOOLS> MAP ALGEBRA> RASTER CALCULATOR. This will open the RASTER CALCULATOR window.

In the RASTER CALCULATOR window, you can build any expression you wish in the lower window. Existing raster data layers can be selected by double clicking on them in the layers and variables section of the window.

2. Build the expression which you want to use to create the new raster data layer

For this exercise, first double click on PA_RASTER to add it to the lower window. Then click on the button with the + sign on it. Next, double click on CD_RASTER_2, followed by the + button, and repeat this for HP_RASTER_2, MW_RASTER_2, WBD_RASTER_2 and TRACK_RASTER. The final expression should read: "PA_RASTER" + "CD_RASTER_2" + "HP_RASTER_2" + "MW_RASTER_2" + "WBD_RASTER_2" + "TRACK_RASTER". **NOTE**: The expression has to be entered exactly like this, including the spaces before and after the plus sign (+). This expression will produce a new raster data layer with values ranging from zero where there has been survey effort but no species were recorded, to five where all five species were recorded. Finally, in the OUTPUT RASTER section of the window type C:\GIS_EXERCISES\SP_RICHNESS and click on the OK button to carry out this calculation.

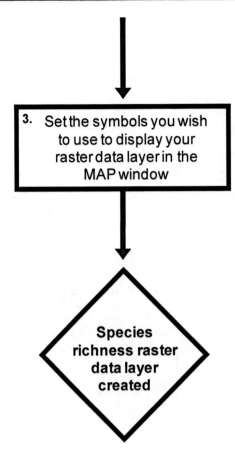

3. Set the symbols you wish to use to display your raster data layer in the MAP window

Right click on the name of your newly created raster data layer (SP_RICHNESS) in the TABLE OF CONTENTS window and select PROPERTIES. **NOTE**: It may have been added below other data layers you already have in your data frame. Next, click on the SYMBOLOGY tab of the LAYER PROPERTIES window. In the left hand portion of the SYMBOLOGY window, select UNIQUE VALUES. Next, click on the ADD ALL VALUES button. Double click on the coloured rectangle beside the number 0 and the second grey box down (which is 10% grey). Next, in the same way select 30% grey for 1, 50% grey for 2, 70% grey for 3, 80% grey for 4 and black for 5. Finally, click the OK button.

Species richness raster data layer created

At the end of step seven, the contents of your MAP window should look like the following image.

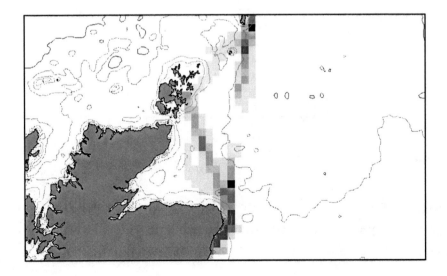

STEP 8: REMOVE UNNEEDED DATA LAYERS FROM YOUR GIS PROJECT:

You will have created a lot of data layers in this exercise, and your TABLE OF CONTENTS window will be looking quite crowded. However, you do not need all of the data layers you created so that you could make your final species richness raster data layer. Therefore, before you finish this exercise, you will remove all these extra data layers from your GIS project. This is done by right clicking on the name of the data layer you wish to remove from your GIS project in the TABLE OF CONTENTS window and selecting REMOVE.

Remove all data layers with the exception of DEPTH_NORTH_SEA, LAND_NORTH_SEA and SP_RICHNESS, so that your TABLE OF CONTENTS looks like this:

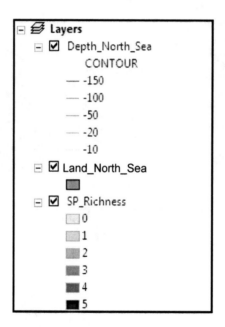

Optional Extra:

In this exercise you created a species richness raster data layer which had values in it which indicated how many species were recorded in each grid cell. However, you could also create a raster data layer where the cell values indicated the total number of cetaceans recorded in each grid cell. To do this, you would need to change the above instructions in three ways. You do not need to do step two, as the ALL_SPECIES data layer already has a field with the number of animals for each sightings in its attribute table. Secondly, once you have selected the data for a particular species from the ALL_SPECIES data layer, you would select the field called NUMBER when creating the species specific raster data layer. You would need to change the rule for what happens when multiple points fall within a single grid cell from AVERAGE to SUM. This involves modifying the instruction set for step five so that it looks like the one provided below.

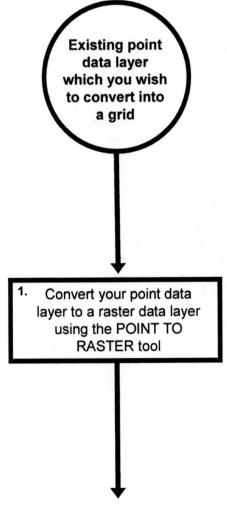

In the toolbox window, select CONVERSION TOOLS> TO RASTER> POINT TO RASTER. In the POINT TO RASTER window, select the point data layer called ALL_SPECIES_PROJECT in the INPUT FEATURES window using the drop down menu. Select the field called NUMBER using the drop down menu in the VALUE FIELD window. Type C:\GIS_EXERCISES\CD_ABUN in the OUTPUT RASTER DATASET window. In the CELL ASSIGNMENT TYPE (OPTIONAL) window select SUM. Type the number 10000 into CELLSIZE (OPTIONAL) window. Next, click on the ENVIRONMENT button at the bottom of the window. In the ENVIRONMENT SETTINGS window, click on PROCESSING EXTENT. In the EXTENT section of the window that appears select SAME AS LAYER TRACK_RASTER from the drop down menu. Now click OK to close the ENVIRONMENTS SETTINGS window. Finally, click OK at the bottom of the POINT TO RASTER window.

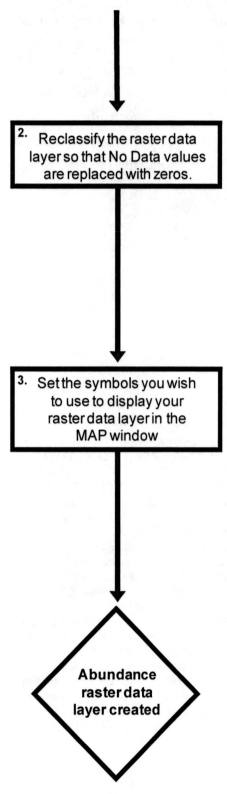

2. Reclassify the raster data layer so that No Data values are replaced with zeros.

3. Set the symbols you wish to use to display your raster data layer in the MAP window

Abundance raster data layer created

In the TOOLBOX window, select SPATIAL ANALYST TOOLS> RECLASS> RECLASSIFY. In the RECLASSIFY window, select ABUN data layer from the drop down menu in the INPUT RASTER window. Select VALUE in the RECLASS FIELD window. In the RECLASSIFICATION window, delete all entries in the RECLASSIFICATION section of the window with the exception of NODATA. This is done by clicking on the grey box at the start of each line and clicking on the DELETE ENTRIES button. Next, type 0 in the NEW VALUES column for the line which has NODATA written in the OLD VALUES column. This will change all the No Data values to zeros. Make sure that there is no tick next to CHANGE MISSING VALUES TO NO DATA (OPTIONAL). Type C:\GIS_EXERCISES\ CD_ABUN_2 in the OUTPUT RASTER window. Now click OK.

Right click on the name of your newly created raster data layer (CD_ABUN_2) in the TABLE OF CONTENTS window and select PROPERTIES. **NOTE**: It may have been added below other data layers you already have in your data frame. Next, click on the SYMBOLOGY tab of the LAYER PROPERTIES window. In the left hand portion of the SYMBOLOGY window, select CLASSIFIED. Next, click on the CLASSIFY button. In the CLASSIFICATION window select MANUAL from the dropdown menu for METHOD. In the BREAK VALUES section of the CLASSIFICATION window, enter 0.5, 5, 10, 25 and 500, and click OK. Double click on the coloured rectangle beside the number 0.5 and select 10% grey. For the other values, in ascending order select 30% grey, 50% grey, 70% grey, 80% grey and black. Finally, click the OK button.

You will also need to make an abundance raster data layer for all five species as you do not already have one for bottlenose dolphin. When making these raster data layers, use the names BND_ABUN_2, CD_ABUN_2, HP_ABUN_2, MW_ABUN_2 and WBD_ABUN_2 for bottlenose dolphin, common dolphin, harbour porpoise, minke whale and white-beaked dolphin respectively.

The only other instructions which would need to be altered would be the section on how to display the contents of the abundance raster data layer you create. Rather than using UNIQUE VALUES for this, use the CLASSIFIED option. Using this option, you can select the classifications you want to use for your symbols by clicking on the CLASSIFY button. This opens the CLASSIFICATION window where you can select MANUAL for the method type and enter 0.5, 10, 25, 50 and 500 as the break values in the right hand section of the window. This breaks the display of the data into the following groups: zero animals recorded per grid cell, 1-10 animals recorded per grid cell, 11-25 animals recorded per grid cell, 26-50 animals recorded per grid cell and 51-500 animals recorded per grid cell. Now click OK to close this window. In the SYMBOLOGY tab of the LAYER PROPERTIES window, you can now select the colours you want to use for each range by clicking on the coloured rectangle next to it. In ascending order use 10% grey, 30% grey, 50% grey, 70% grey, 80% grey and black for the five ranges. Finally, click OK to close the LAYER PROPERTIES window. Once you have finished creating your cetacean abundance raster data layer, the contents of your TABLE OF CONTENTS window should look like this:

And the contents of your MAP window should look like this.

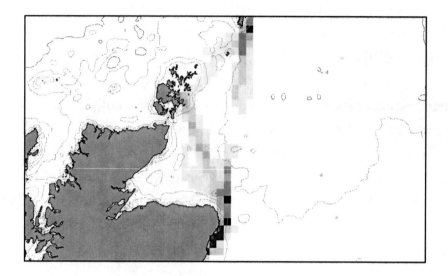

Chapter Six

Exercise Four: Creating A Polygon Grid Data Layer Of Abundance Per Unit Effort From Survey Data

In exercise two, you produced a raster data layer which showed where bottlenose dolphin were recorded in relation to those areas which were surveyed. However, this only contained information about species presence and absence. While this is useful for some research questions, for others you may need to know how many individuals of a species were recorded in each grid cell, and more importantly, how many were recorded per unit of survey effort in each grid cell rather than simply whether the species was present or not. This abundance per unit of survey effort could be the number of fish of a particular species caught per distance trawled from trawl surveys, or recorded abundance of dolphins of a particular species per distance surveyed from a cetacean sightings survey. Both can be calculated in a similar manner.

Rather than creating a raster data layer as you did when creating a presence-absence grid (exercise two) and a species richness grid (exercise three), it is easier to calculate abundance per unit effort for individual grid cells using something known as a polygon grid. A polygon grid is a polygon data layer where the polygons are squares that represent non-overlapping grid cells which form a continuous surface across a specific area. Polygon grid data layers have the advantage over grids represented as raster data layers in that they have attribute tables associated with them. This means that while a raster data layer can usually only contain information about a single variable, the attribute table of a polygon grid can contain information about a number of variables in different fields. In addition, as it is a feature data layer, it can interact with other feature data layers in ways that raster data layers cannot. This provides greater flexibility when carrying out a number of different tasks, and its use is critical to one step in the process of calculating abundance per unit effort. This is step seven where the polygon grid is used to divide up the survey tracks into sections which fall within each individual grid cell so that the total amount of survey effort per grid cell can be calculated. This would not be possible to do with a grid represented as a raster data layer. In this exercise, you will create a polygon grid data layer and use it to calculate the number of bottlenose dolphins recorded per kilometre of survey effort in each grid cell for the surveys from northeast Scotland. This is the longest and most complicated exercise in this book, and you may find that you need to go back and repeat some of the steps at some stage if you go wrong. You can make it easier to do this if, at the end of each step, you save your GIS project under a slightly different name. For example, at the start of this exercise, your GIS project will be saved as EXERCISE_FOUR. When you complete step one, select SAVE AS from the FILE menu in the main menu bar area, and save it as EXERCISE_FOUR_STEP_ONE. Do the same at the end of every other step, updating the number at the end accordingly. This means that you can always go back a step and start again if you make a mistake in a particular step. It is also a good idea to do this on a regular basis when using GIS within your own research.

However, you may find that if you have edited any of your data layers (for example adding a field to the attribute table, or filling it in with the FIELD CALCULATOR or CALCULATE GEOMETRY tools), that you have to delete these new fields from the data layer before you can start a specific step again. This is because these changes will have been made to the actual source file of the data layer and not just to the contents of the GIS project file. A field can be deleted by right-clicking on the field's name in the TABLE window and selecting DELETE. However, this action is permanent, so make sure that you select the right field and that you definitely want to delete it before doing so. If you find that you get completely lost and cannot get a data layer back to the way you need it for the start of a step you wish to repeat. You can remove it from your GIS project (by right-clicking on its name in the TABLE OF CONTENTS window and selecting REMOVE) and either download it again from the website (if it was one of the original data layers provided), or you can make it again (if it was one of the data layers made during this or a previous exercise). **NOTE**: If you do repeat a step, you may find that data layers with the names used in the instructions already exist, and that you will be asked whether you want to over-write them. This is OK as you will not want to keep data layers which have been wrongly made. However, be careful about over-writing data layers when doing your own research as this will permanently delete the original data layer.

For this exercise, you will use four existing data layers. These data layers should all be saved in a folder on your C drive called GIS_EXERCISES prior to starting this exercise, so that it has the address C:\GIS_EXERCISES\. These data layers are:

1. All_Species.shp: This is the point data layer of locational records for all species recorded during the surveys. Each point represents a record for a single group of cetaceans recorded during a set of surveys. It is in the geographic projection and is based on the WGS 1984 datum.

2. Survey_Tracks_North_Sea.shp: This is a line data layer which contains information on survey effort for the study which recorded the species locational data in All_Species.shp. Thus, this data layer tells you which locations were surveyed during the study from which the sightings data come. It is in the geographic projection and is based on the WGS 1984 datum.

3. Depth_North_Sea.shp: This is a line data layer which contains information on water depth within the study area. **NOTE**: The depth values are not accurate and are only provided here as an approximation for training purposes. As a result, it should not be used for any other purpose. This data layer is in the geographic projection and is based on the WGS 1984 datum. This data layer is included in this GIS project is for illustrative purposes only.

4. Land_North_Sea.shp: This is a polygon data layer which contains information on land in this region which you can use to create your map. **NOTE**: The land information is not accurate and is only provided here as an approximation for illustration purposes only. As a result, it should not be used for any other purpose. This data layer is in the geographic projection and is based on the WGS 1984 datum.

Once you have all these files downloaded into the correct folder on your computer, and understand what is contained within each file, you can move onto creating your map. The starting point for this is a blank GIS project. To create a blank GIS project, first, start the ArcGIS 10.1 software by opening the ArcMap module. When it opens, you will be presented with a window which has the heading ARCMAP – GETTING STARTED. In this window you can either select an existing GIS project to work on, or create a new one. To create a new blank GIS project, click on NEW MAPS in the directory tree on the left hand side and then select BLANK MAP in the right hand section of the window. Now, click OPEN at the bottom of this window. Once you have opened your new GIS project, the first thing you need to do is save it under a specific name. To do this, click on FILE from the main menu bar area, and select SAVE AS. For this example, save it as EXERCISE_FOUR in C:\GIS_EXERCISES\.

Summary Flow Diagram For Creating A Polygon Grid Data Layer With Abundance Per Unit Effort From Survey Data

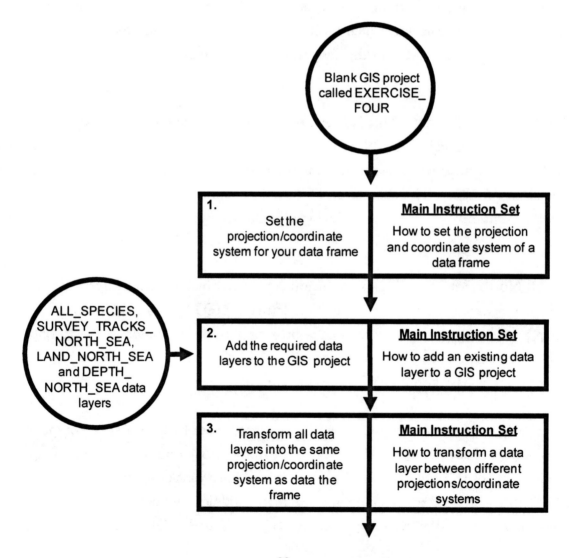

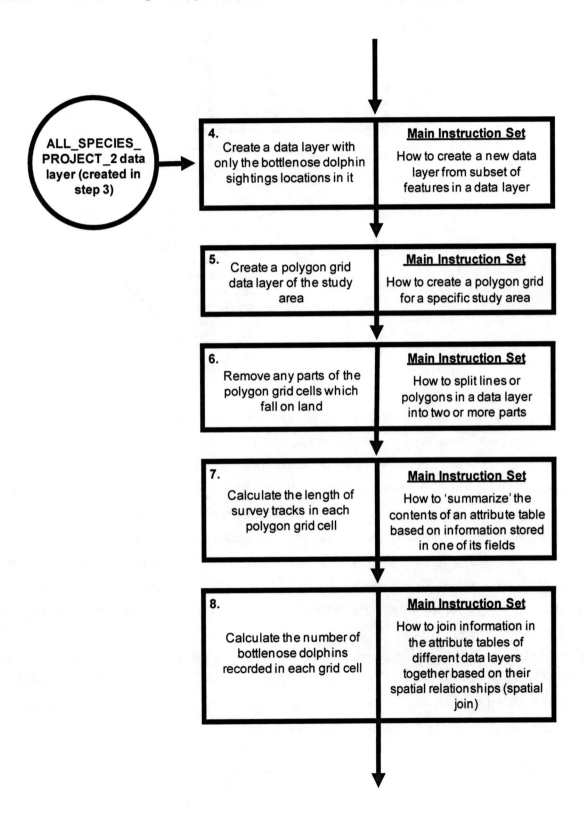

ALL_SPECIES_
PROJECT_2 data
layer (created in
step 3)

4.
Create a data layer with
only the bottlenose dolphin
sightings locations in it

Main Instruction Set

How to create a new data
layer from subset of
features in a data layer

5.
Create a polygon grid
data layer of the study
area

Main Instruction Set

How to create a polygon grid
for a specific study area

6.
Remove any parts of the
polygon grid cells which
fall on land

Main Instruction Set

How to split lines or
polygons in a data layer
into two or more parts

7.
Calculate the length of
survey tracks in each
polygon grid cell

Main Instruction Set

How to 'summarize' the
contents of an attribute table
based on information stored
in one of its fields

8.
Calculate the number of
bottlenose dolphins
recorded in each grid cell

Main Instruction Set

How to join information in
the attribute tables of
different data layers
together based on their
spatial relationships (spatial
join)

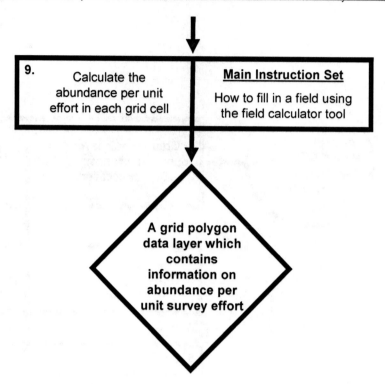

Once you have familiarised yourself with the summary flow diagram, then you need to read through the instruction set for the first step, which can be found below. Once you have completed the first step, read through the instructions again and make sure that you have completed it properly. It is important to do this at this stage as you need to use the results of one step as the starting point for the next, for this reason it is important to make sure that you have completed it correctly at this stage. In general, it is much easier to spot where you have gone wrong at the end of an individual step, rather than trying to work it out later. Once you have successfully completed the first step, move onto the second step and repeat this process, and so on until you have completed all the steps in the summary flow diagram.

Within the instructions for each step, images of the contents of the MAP window, the LAYOUT window, the TABLE OF CONTENTS window and/or the TABLE window will be provided so that you have an idea of what your GIS project should look like at that specific point as you progress through this exercise.

Instruction Sets For The Major Steps Identified In The Summary Flow Diagram:

STEP 1: SET THE PROJECTION/COORDINATE SYSTEM FOR YOUR DATA FRAME:

This will use the instruction sets *How to set the projection and coordinate system for a data frame in a GIS project* (from *An Introduction To Using GIS In Marine Biology*). For this exercise, the projection/coordinate system that will be used is the custom transverse mercator centred on latitude 56.5°N and longitude 1.0°W (or latitude 56.5° and longitude -1.5° in decimal degrees) based on the WGS 1984 datum introduced in exercise one. This is the projection/coordinate system which you will set your data frame to in step one.

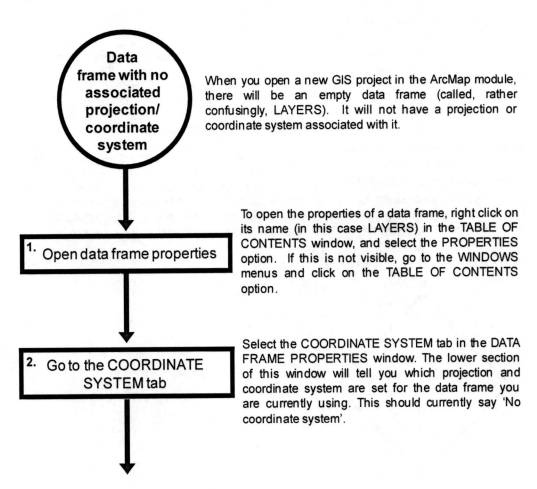

Data frame with no associated projection/ coordinate system

When you open a new GIS project in the ArcMap module, there will be an empty data frame (called, rather confusingly, LAYERS). It will not have a projection or coordinate system associated with it.

1. Open data frame properties

To open the properties of a data frame, right click on its name (in this case LAYERS) in the TABLE OF CONTENTS window, and select the PROPERTIES option. If this is not visible, go to the WINDOWS menus and click on the TABLE OF CONTENTS option.

2. Go to the COORDINATE SYSTEM tab

Select the COORDINATE SYSTEM tab in the DATA FRAME PROPERTIES window. The lower section of this window will tell you which projection and coordinate system are set for the data frame you are currently using. This should currently say 'No coordinate system'.

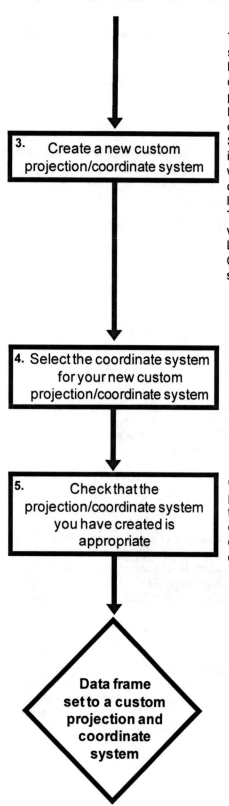

3. Create a new custom projection/coordinate system

4. Select the coordinate system for your new custom projection/coordinate system

5. Check that the projection/coordinate system you have created is appropriate

Data frame set to a custom projection and coordinate system

To create a new custom projection/coordinate system, press the ADD COORDINATE SYSTEM button that can be found towards the top right hand corner of the COORDINATE SYSTEM tab (it has a picture of a globe on it). and select NEW> PROJECTED COORDINATE SYSTEM. This will open the NEW PROJECTED COORDINATE SYSTEM window. In the upper NAME window, type in NORTH SEA. In the PROJECTION portion of the window, select the name of the appropriate type of coordinate system from the drop down menu (in the lower NAME window). For this exercise, select TRANSVERSE MERCATOR. Next type in the values you wish to use for the parameters. LATITUDE_OF_ORIGIN enter 56.5. For CENTRAL_MERIDIAN enter -1.0. Leave all other sections of the window with their default settings.

In the GEOGRAPHIC COORDINATE SYSTEM section of the PROJECTED COORDINATE SYSTEM window, by default it should say NAME: GCS_WGS_1984. If is doesn't, click on the CHANGE button and type WGS 1984 into the SEARCH box in the window that appears and press the return key on your keyboard. Select WORLD> WGS 1984, and click the OK button. Now click the OK button in the NEW PROJECTED COORDINATE SYSTEM window. Finally, click OK in the DATA FRAME PROPERTIES window.

Once you have created a custom projection/coordinate system, you need to check that it is appropriate. This involves examining how data layers look in it. For this exercise, this will be done in the next step by adding a polygon data layer of land and checking that it looks the right shape.

To check that you have done this step properly, right click on the name of your data frame (LAYERS) in the TABLE OF CONTENTS window and select properties. Click on the COORDINATE SYSTEM tab and make sure that the contents of the CURRENT COORDINATE SYSTEM section of the window has the following text at the top of it:

>North Sea
>Authority: Custom
>
>Projection: Transverse_Mercator
>False_Easting: 0.0
>False_Northing: 0.0
>Central_Meridian: -1.0
>Scale_Factor: 1.0
>Latitude_Of_Origin: 56.5
>Linear Unit: Meter (1.0)

If it does not, you will need to repeat this step to ensure that you have assigned the correct projection/coordinate system to your data frame.

Now click OK to close the DATA FRAME PROPERTIES window.

Finally, for this step, you will set the extent of your data frame to the same fixed extent that has been used in previous exercises. This is not something you would normally do in a GIS project as it means that you cannot zoom in or out, but it means that the extent of your MAP window will be fixed, making comparisons with the figures of how the contents of your MAP window should look at various points throughout this exercise much easier. To set the extent of your data frame, right click on its name in the TABLE OF CONTENTS window and select PROPERTIES. In the DATA FRAME PROPERTIES window, select the DATA FRAME tab. In the DATA FRAME tab, select FIXED EXTENT and type in the following coordinates: TOP: 388000; LEFT: -130000; RIGHT: 50000; BOTTOM: 56000. Once these have been entered, click OK. Your data frame will now be set to this fixed extent, and the ZOOM and PAN tools will have been disabled.

STEP 2: ADD THE REQUIRED DATA LAYERS TO THE GIS PROJECT:

Once the projection/coordinate system has been set for your data frame, you need to add the data layers which you will use for this exercise. These are ALL_SPECIES.SHP, SURVEY_TRACKS_NORTH_SEA.SHP, LAND_NORTH_SEA.SHP and DEPTH_ NORTH_SEA.SHP. **NOTE**: The data layer DEPTH_NORTH_SEA will not be directly used in this exercise, but is added to provide additional information for display purposes.

The instructions for this step are based on instruction sets called *How to add an existing data layer to a GIS project* and *How to change the display symbols for a data layer* (from *An Introduction To Using GIS In Marine Biology*).

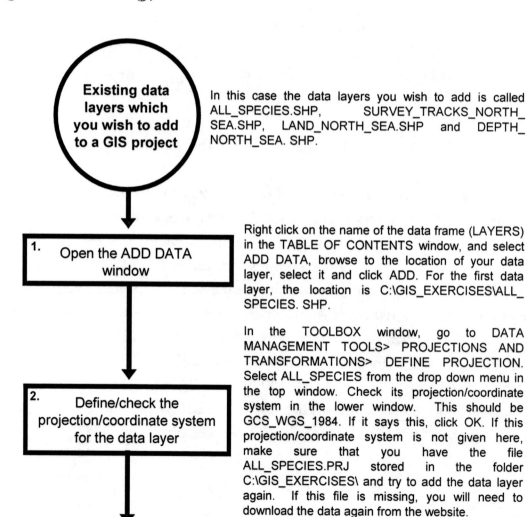

Existing data layers which you wish to add to a GIS project

In this case the data layers you wish to add is called ALL_SPECIES.SHP, SURVEY_TRACKS_NORTH_ SEA.SHP, LAND_NORTH_SEA.SHP and DEPTH_ NORTH_SEA. SHP.

1. Open the ADD DATA window

Right click on the name of the data frame (LAYERS) in the TABLE OF CONTENTS window, and select ADD DATA, browse to the location of your data layer, select it and click ADD. For the first data layer, the location is C:\GIS_EXERCISES\ALL_ SPECIES. SHP.

2. Define/check the projection/coordinate system for the data layer

In the TOOLBOX window, go to DATA MANAGEMENT TOOLS> PROJECTIONS AND TRANSFORMATIONS> DEFINE PROJECTION. Select ALL_SPECIES from the drop down menu in the top window. Check its projection/coordinate system in the lower window. This should be GCS_WGS_1984. If it says this, click OK. If this projection/coordinate system is not given here, make sure that you have the file ALL_SPECIES.PRJ stored in the folder C:\GIS_EXERCISES\ and try to add the data layer again. If this file is missing, you will need to download the data again from the website.

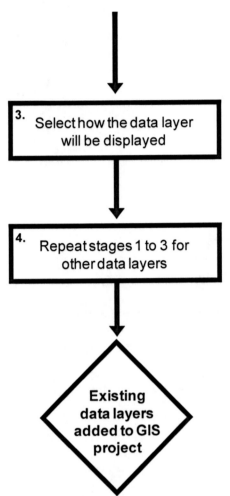

Right click on the name ALL_SPECIES in the TABLE OF CONTENTS window, and select PROPERTIES. Next, click on the SYMBOLOGY tab. Now set the ALL_SPECIES data layer to display in the same way as when it was used in exercise three (see page 62).

Repeat stages one to three of this instruction set for SURVEY_TRACKS_NORTH_SEA.SHP, LAND_ NORTH_SEA.SHP and DEPTH_NORTH_SEA. SHP. When selecting how these data layers should be displayed, for DEPTH_NORTH_SEA, use the legend provided on page 20. For SURVEY_ TRACKS_NORTH_SEA use a grey line of size 3, and for LAND_NORTH_SEA use dark green as the fill colour.

Once you have completed this step, you may need to re-order the data layers so that ALL_SPECIES is at the top of the TABLE OF CONTENTS window, followed by SURVEY_TRACKS_NORTH_SEA, DEPTH_NORTH_SEA and LAND_NORTH_ SEA. This can be done by clicking on a data layers name in the TABLE OF CONTENTS window and holding the mouse button down while dragging a data layer upwards or downwards to its desired new position.

At this point, your TABLE OF CONTENTS window should look like the figure at the top of the next page.

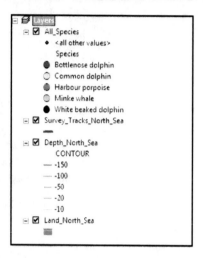

While the contents of your MAP window should look like this:

If the contents of your MAP window do not look like this, look in the TABLE OF CONTENTS window and check that you have all the required data layers added to your GIS project and that they are in the right order. Secondly, check that you have the data frame set to the correct extent by right-clicking on the data frame's name in the TABLE OF CONTENTS window and selecting PROPERTIES. In the DATA FRAME PROPERTIES window, click on the DATA FRAME tab and make sure that the extent is set to the FIXED EXTENT given at the end of step one.

STEP 3: TRANSFORM ALL DATA LAYERS INTO THE SAME PROJECTION/COORDINATE SYSTEM AS THE DATA FRAME:

At this stage, the contents of your MAP window will look very similar to how it looked at the end of exercise one, with the exception that you have locational records for more than one species and you also now have survey track data added to your GIS project. However, while you will not be able to see it by looking at the contents of your MAP window, your data layers are in one projection/coordinate system (the geographic projection), and your data frame is in another (the custom transverse mercator projection called North Sea). This is not necessarily an issue when creating maps and figures, but if you actually want to start comparing and using your data layers to create new information based on the relative spatial distributions of features within them, it is usually best to ensure that the data layers and the data frame are all in the same projection/coordinate system. This involves either transforming the data layers to the same projection/coordinate system of the data frame, or changing the projection/coordinate system of the data frame to that of the data layer.

In this exercise, you will need to measure the lengths of the survey tracks in order to be able to calculate the number of bottlenose dolphins per unit effort in each grid cell of a polygon grid. Therefore, you need to use a projection/coordinate system which is suitable for doing this. A geographic projection is not, while a transverse mercator projection centred on the middle of the study area is. As a result, you will transform the data layers from their current geographic projection to the same custom transverse mercator projection as the data frame.

When you transform your data layers into this projection/coordinate system, you will find that you need to create the custom transverse mercator projection/coordinate system from scratch. In this exercise this will be repeated for each data layer in order to keep things as simple as possible. However, after you have transformed the first data layer, it is possible to import the projection/coordinate system from it when transforming the other data layers, rather than having to make it from scratch again each time. This is done by selecting IMPORT rather than NEW in the SPATIAL REFERENCE PROPERTIES window when transforming your data layer from one projection/coordinate system to another and selecting a data layer which is already in the appropriate projection/coordinate system.

These instructions as based on instruction sets called *How to transform data layers between different projections* and *How to change the display symbols for a data layer* from *An Introduction To Using GIS In Marine Biology*.

Existing data layers which you wish to transform into a new projection/ coordinate system

In this case the data layers you wish to transform are ALL_SPECIES, SURVEY_TRACKS_NORTH_SEA, LAND_NORTH_SEA and DEPTH_ NORTH_SEA. You will start with the data layer called ALL_SPECIES.

1. Transform your data layer to the same projection/ coordinate system as the data frame

In the TOOLBOX window, select DATA MANAGEMENT TOOLS> PROJECTIONS AND TRANSFORMATIONS> FEATURE> PROJECT. This will open the PROJECT window. Select ALL_SPECIES from the drop down menu in the INPUT DATASET OR FEATURE CLASS section of the window. In the OUTPUT DATASET OR FEATURE CLASS section enter C:\GIS_EXERCISES\ALL_SPECIES_ PROJECT_2.

You will need to create the custom projection/coordinate system you are going to transform the data layers into. This is done by clicking on the button at the end of the OUTPUT COORDINATE SYSTEM section of the window to open the SPATIAL REFERENCE PROPERTIES window. Next, click on the ADD COORDINATE SYSTEM button in the top right hand corner of the SPATIAL REFERENCE PROPERTIES window and select NEW> PROJECTED COORDINATE SYSTEM. In the NEW PROJECTED COORDINATE SYSTEM window, type in NORTH SEA into the NAME section. In the PROJECTION section of the window, select TRANSVERSE MERCATOR from the drop down menu. Next type in 56.5 for LATITUDE_OF_ORIGIN and -1.0 for CENTRAL_MERIDIAN. Leave all other fields with their default settings.

In the GEOGRAPHIC COORDINATE SYSTEM section of the NEW PROJECTED COORDINATE SYSTEM window, by default it should say NAME: GCS_WGS_1984. If is doesn't, click on the CHANGE button and type WGS 1984 into the SEARCH box in the window that appears and press the return key on your keyboard. Select WORLD> WGS 1984, and click the OK button. Now click the OK button in the NEW PROJECTED COORDINATE SYSTEM window. Next click the OK button in the SPATIAL REFERENCE PROPERTIES window. Finally, click OK in the PROJECT window.

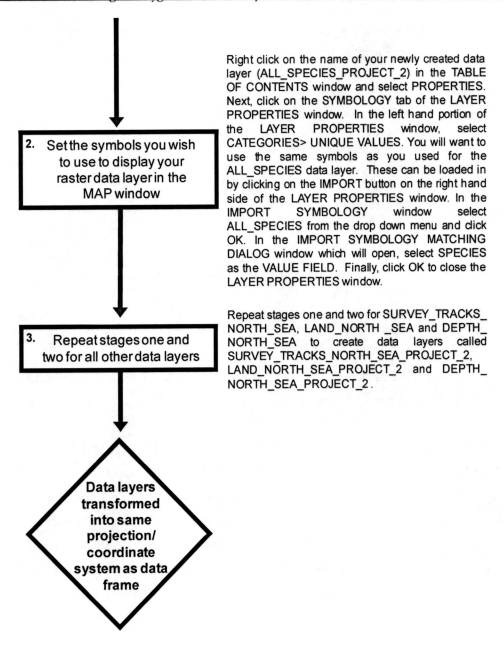

2. Set the symbols you wish to use to display your raster data layer in the MAP window

Right click on the name of your newly created data layer (ALL_SPECIES_PROJECT_2) in the TABLE OF CONTENTS window and select PROPERTIES. Next, click on the SYMBOLOGY tab of the LAYER PROPERTIES window. In the left hand portion of the LAYER PROPERTIES window, select CATEGORIES> UNIQUE VALUES. You will want to use the same symbols as you used for the ALL_SPECIES data layer. These can be loaded in by clicking on the IMPORT button on the right hand side of the LAYER PROPERTIES window. In the IMPORT SYMBOLOGY window select ALL_SPECIES from the drop down menu and click OK. In the IMPORT SYMBOLOGY MATCHING DIALOG window which will open, select SPECIES as the VALUE FIELD. Finally, click OK to close the LAYER PROPERTIES window.

3. Repeat stages one and two for all other data layers

Repeat stages one and two for SURVEY_TRACKS_NORTH_SEA, LAND_NORTH _SEA and DEPTH_NORTH_SEA to create data layers called SURVEY_TRACKS_NORTH_SEA_PROJECT_2, LAND_NORTH_SEA_PROJECT_2 and DEPTH_NORTH_SEA_PROJECT_2.

Data layers transformed into same projection/ coordinate system as data frame

To check that you have done this step properly, right click on the name of each new data layer (e.g. ALL_SPECIES_PROJECT_2) in the TABLE OF CONTENTS window and select properties. Click on the SOURCE tab and make sure that the contents of the DATA

SOURCE section of the window has the following text towards the bottom of it (you may have to scroll down to see it all):

Projected Coordinate System: NORTH SEA
Projection: Transverse_Mercator
False_Easting: 0.00000000
False_Northing: 0.00000000
Central_Meridian: -1.00000000
Scale_Factor: 1.00000000
Latitude_Of_Origin: 56.50000000
Linear Unit: Meter

Geographic Coordinate System: GCS_WGS_1984
Datum: D_WGS_1984
Prime Meridian: Greenwich
Angular Unit: Degree

If it does not, you will need to repeat this step to ensure that you have assigned the correct projection/coordinate system to your data frame.

Next, remove the data layers ALL_SPECIES, SURVEY_TRACKS_NORTH_SEA LAND_NORTH_SEA and DEPTH_NORTH_SEA from your GIS project by right-clicking on their name in the TABLE OF CONTENTS window and selecting REMOVE. Finally, re-arrange the order of the remaining data layers (by clicking on a data layer's name in the TABLE OF CONTENTS window and, while holding down the mouse button, dragging it up or down until it is in the desired position) so that your TABLE OF CONTENTS window looks like the one at the top of the next page.

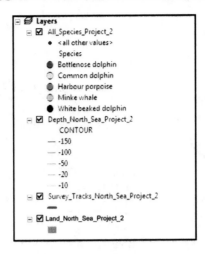

The contents of your MAP window should look the same as before. This is because you have the same information in all your data layers, you have simply transformed the projection/coordinate system of the data layers to a different one.

STEP 4: CREATE A DATA LAYER WITH ONLY THE BOTTLENOSE DOLPHIN SIGHTINGS LOCATIONS IN IT:

In exercise one, you created a data layer of bottlenose dolphin sightings locations from latitude and longitude data. In this step, you will create a data layer containing the same information, but in a different way. This is because it demonstrates how you can create new data layers from subsets of features in existing data layers. Therefore, in this step, you will create a data layer of bottlenose dolphin sightings locations by selecting just the sightings for this species from the ALL_SPECIES_PROJECT_2 data layer using the SELECT BY ATTRIBUTE tool and then exporting these features to a new data layer.

These instructions are based on instruction sets called *How to make a new data layer from a subset of features in an existing data layer* and *How to change the display symbols for a data layer* from *An Introduction To Using GIS In Marine Biology.*

Existing data layer from which you wish to make a new data layer based on a subset of its features

In this exercise, this data layer is ALL_SPECIES_ PROJECT_2.

Click on SELECTION in the main menu bar area and select SELECT BY ATTRIBUTES. In the SELECT BY ATTRIBUTES window, select ALL_SPECIES_PROJECT_2 from the drop down menu in the LAYER section of the window. Select CREATE A NEW SELECTION from the METHOD section of the window. Next, in the section below that, select SPECIES by double clicking on its name. It will now appear in the bottom section of the window.

Now, click on the button with the equals sign (=) on it. This will add it to the expression being built in the bottom section of the window. Next, click on the GET UNIQUE VALUES button. This will bring up a list of the five species names found in the SPECIES field directly above this button. You can select the species name you want to select by double clicking on it to add it to the expression. Double click on BOTTLENOSE DOLPHIN. The expression in the lower section of the SELECT BY ATTRIBUTES window should now read: "Species" = 'Bottlenose dolphin'. Now click OK.

1. Select the feature or features using the SELECT BY ATTRIBUTES tool

Finally, you need to check that all the bottlenose dolphin records and no data for other species have been selected. Right click on the name ALL_SPECIES_PROJECT_2 in the TABLE OF CONTENTS window and select OPEN ATTRIBUTE TABLE. Scroll down in this table and make sure only lines with BOTTLENOSE DOLPHIN in the SPECIES field have been selected. Once this has been checked, close the ATTRIBUTE TABLE window.

2. Make a new data layer based on these selected features

To make a new data layer from the selected records, right click on the data layer's name (ALL_SPECIES_PROJECT_2) in the TABLE OF CONTENTS window. Select DATA> EXPORT DATA. In the EXPORT DATA window, make sure that next to EXPORT you select SELECTED FEATURES. In the OUTPUT FEATURE CLASS section of the EXPORT DATA window type C:\GIS_EXERCISES\BOTTLENOSE_DOLPHIN_ PROJECT_2. When asked 'Do you want to add the exported data to the map as a layer' click YES.

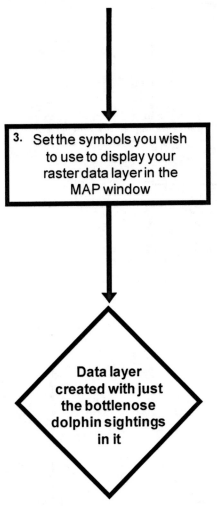

3. Set the symbols you wish to use to display your raster data layer in the MAP window

Right click on the name of your newly created data layer (BOTTLENOSE_DOLPHIN_ PROJECT_2) in the TABLE OF CONTENTS window and select PROPERTIES. Next, click on the SYMBOLOGY tab of the LAYER PROPERTIES window. In the left hand portion of the LAYER PROPERTIES window, select FEATURES> SINGLE SYMBOL. Next, click on the button with the coloured circle on it to open the SYMBOL SELECTOR window. Select CIRCLE 2, size 12 and colour PONSIETTA RED. Click the OK button to close the SYMBOL SELECTOR window. Finally, click OK to close the LAYER PROPERTIES window.

Data layer created with just the bottlenose dolphin sightings in it

At the end of this step, remove the data layer ALL_SPECIES_PROJECT_2 by right-clicking on its name in the TABLE OF CONTENTS window and selecting REMOVE.

Your TABLE OF CONTENTS window should now look like the figure at the top of the next page.

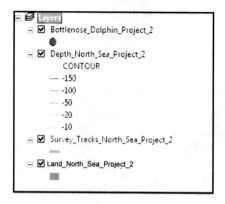

While the contents of your MAP window should look like this:

STEP 5: CREATE A POLYGON GRID DATA LAYER FOR THE STUDY AREA:

In order to calculate the recorded abundance of bottlenose dolphins per grid cell in a polygon grid data layer, you first need to create a polygon grid data layer and this will be created in this step.

In versions of ArcGIS prior to 10.1, the process of creating a polygon grid data layer was not necessarily as straight-forward as you would think it should be. Specifically, it involved creating a raster data layer, converting it into a point data layer before converting it back into a raster data layer and then converting it into the final polygon data layer. This needed to be done in this way because you can only convert a raster data layer into a polygon data layer if it contains integer values (i.e. none of the values contain decimal values), and each grid cell value must be unique. The quickest and easiest way to generate a raster data layer which has a unique integer value for each grid cell was to convert a raster data layer into a point data layer, and then convert it back to a raster data layer based not on the field which has the value from the original raster data layer, but using the unique FID (or feature ID) field which was created when the point data layer was made.

Now, using ArcGIS 10.1, you can create a polygon grid data layer using the CREATE FISHNET tool (DATA MANAGEMENT TOOLS> FEATURE CLASS> CREATE FISHNET – see *How to create a polygon grid for a specific study area* from *An Introduction To Using GIS In Marine Biology* for more information on using this tool). However, for this exercise, you will use the approach developed for the older versions of the software because it provides you with experience in converting between different types of data layers.

You will make your polygon grid so that it directly matches up with your presence-absence and species richness raster data layers. Therefore, you will use the same cell size, extent and projection/coordinate system as these raster data layers.

These instructions are based on *How to create a polygon grid for a specific study area* and *How to change the display symbols for a data layer* from *An Introduction To Using GIS In Marine Biology*.

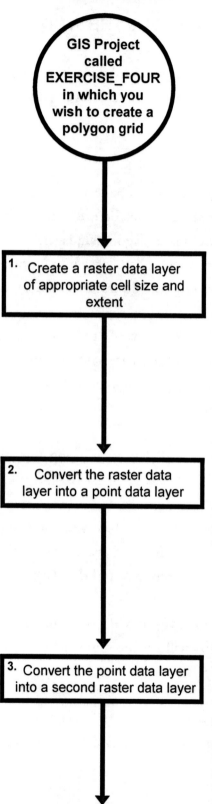

GIS Project called EXERCISE_FOUR in which you wish to create a polygon grid

1. Create a raster data layer of appropriate cell size and extent

2. Convert the raster data layer into a point data layer

3. Convert the point data layer into a second raster data layer

In the TOOLBOX window, select DATA MANAGEMENT TOOLS> RASTER> RASTER DATASET> CREATE RANDOM RASTER. When the CREATE RANDOM RASTER window opens, type C:\GIS_EXERCISES in the OUTPUT LOCATION window. Type the name RANDOM_RASTER in the RASTER DATASET NAME WITH EXTENSION section of the window. Type 10000 in the CELL SIZE (OPTIONAL) section (you may need to scroll down to find it). Finally, type 396000 for TOP, -130000 for LEFT, 50000 for RIGHT and 56000 for BOTTOM. Now click on the ENVIRONMENTS button and select OUTPUT COORDINATES. From the options that appear select SAME AS DISPLAY from the drop down menu. Now click OK to close the ENVIRONMENT SETTINGS window. Finally, click OK to close the CREATE RANDOM RASTER window.

In the TOOLBOX window, select CONVERSION TOOLS> FROM RASTER> RASTER TO POINT. Select the raster you have just created (RANDOM_RASTER) from the drop down menu in the INPUT RASTER section of the window. Type C:\GIS_EXERCISES\RANDOM_POINTS in the OUTPUT POINT FEATURES section of the window. Now click OK. This creates a point data layer with the point placed at the centre of each grid cell.

In the TOOLBOX window, select CONVERSION TOOLS> TO RASTER> POINT TO RASTER. Select RANDOM_POINTS in the INPUT FEATURES section of the window. In the VALUE FIELD section, select POINTID. In the OUTPUT RASTER DATASET section, type C:\GIS_EXERCISES\POINT_RASTER. Type 10000 into the CELL SIZE (OPTIONAL) window. Now click on the ENVIRONMENTS button at the bottom of the POINT TO RASTER window. In the ENVIRONMENT SETTINGS window, click on PROCESSING EXTENT. From the options that appear, select SAME AS LAYER RANDOM_RASTER from the drop down menu. Now click OK to close the ENVIRONMENT SETTINGS window, and then OK to close the POINT TO RASTER window.

106

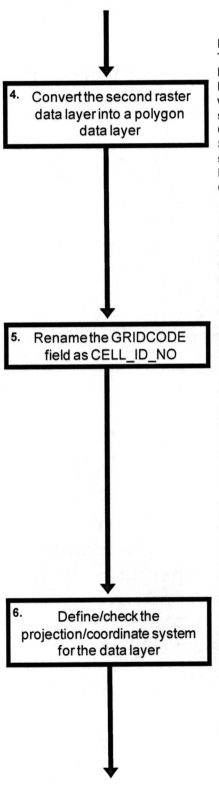

4. Convert the second raster data layer into a polygon data layer

5. Rename the GRIDCODE field as CELL_ID_NO

6. Define/check the projection/coordinate system for the data layer

In the TOOLBOX window, select CONVERSION TOOLS> FROM RASTER> RASTER TO POLYGON. Select POINT_RASTER in the INPUT RASTER section of the RASTER TO POLYGON window using the drop down menu. In the FIELD section of the window select VALUE. Type C:\GIS_EXERCISES\POLYGON_GRID_NORTH_ SEA into the OUTPUT POLYGON FEATURES section. Next, make sure that the SIMPLIFY POLYGONS OPTIONAL is <u>NOT</u> ticked. Finally, click OK.

Right click on the name POLYGON_ GRID_NORTH_SEA in the TABLE OF CONTENTS window. Select OPEN ATTRIBUTE TABLE. Click on the OPTIONS at the top left of the TABLE window, and select ADD FIELD. In the ADD FIELD window, under NAME type 'CELL_ID_NO'. For PRECISION type in 16. Under TYPE select LONG INTEGER. Now click OK. Next right click on the name of the new field that you have just added (CELL_ID_NO) and select FIELD CALCULATOR. When the FIELD CALCULATOR window opens, if it asks if you wish to continue click the YES button. In the FIELDS window, double click on GRIDCODE. This should then appear in the lower window. Now click OK. This transfers the unique ID numbers from the field called GRIDCODE to the new field called CELL_ID_NO. Now you need to delete the column called GRIDCODE. This is done by right-clicking on the field name GRIDCODE and selecting DELETE FIELD.

In the TOOLBOX window, go to DATA MANAGEMENT TOOLS> PROJECTIONS AND TRANSFORMATIONS> DEFINE PROJECTION. Select POLYGON_GRID_NORTH_SEA from the drop down menu in the top section of the window. Check its projection/coordinate system in the lower window. This should be NORTH SEA. If so, click OK to close the DEFINE PROJECTION window. If this projection/coordinate system is not given here, click on the button at the end of the COORDINATE section of the window. In the SPATIAL REFERENCE PROPERTIES window, click on the ADD COORDINATE SYSTEM button and select IMPORT. In the window that opens, select BOTTLENOSE_DOLPHIN_PROJECT_2 (in C:/GIS _EXERCISES/). Next, click ADD to close this window and then OK to close the SPATIAL REFERENCE PROPERTIES window. Finally, click OK in the DEFINE PROJECTION window.

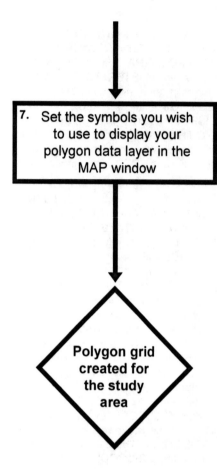

7. Set the symbols you wish to use to display your polygon data layer in the MAP window

Right click on the name of your newly created polygon data layer (POLYGON_GRID_NORTH_ SEA) in the TABLE OF CONTENTS window and select PROPERTIES. Next, click on the SYMBOLOGY tab of the LAYER PROPERTIES window. In the left hand portion of the LAYER PROPERTIES window, select FEATURES> SINGLE SYMBOL. Next, click on the button with the coloured rectangle on it to open the SYMBOL SELECTOR window. Select HOLLOW in the left hand section of the window. Click the OK button to close the SYMBOL SELECTOR window. Finally, click OK to close the LAYER PROPERTIES window.

Polygon grid created for the study area

At the end of this step, remove the data layers RANDOM_RASTER, RANDOM_POINTS and POINT_RASTER by right-clicking on their name in the TABLE OF CONTENTS window and selecting REMOVE.

At the end of this step, the contents of your MAP window should look like the figure at the top of the next page.

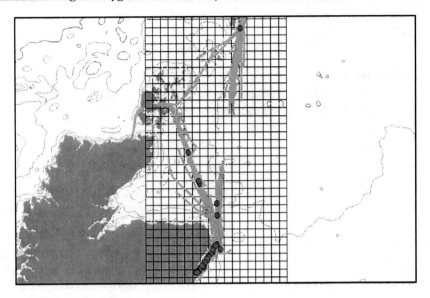

STEP 6: REMOVE ANY PARTS OF THE POLYGON GRID CELLS WHICH FALL ON LAND:

At the moment, your polygon grid data layer for the study area also contains grid cells which lie either partially or completely on land. It is useful to be able to remove grid cells which are completely on land, and also to remove the land sections of ones which are partially on land. This will be done in this step and requires that you have a polygon data layer of the land within the study area (in this case LAND_NORTH_SEA_PROJECT_2.

These instructions are based on the instruction sets called *How to split lines or polygons in a data layer into two or more parts, How to remove features from an existing data layer* and *How to change the display symbols for a data layer.* The generic versions of these instruction sets can be found in *An Introduction To Using GIS In Marine Biology.*

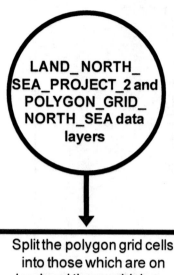

LAND_NORTH_ SEA_PROJECT_2 and POLYGON_GRID_ NORTH_SEA data layers

These data layers were created in steps three and five, respectively, of this current exercise.

1. Split the polygon grid cells into those which are on land and those which are over the sea

In the TOOLBOX window, select ANALYSIS TOOLS> OVERLAY> UNION. This will open the UNION tool window. In the UNION tool window, select POLYGON_GRID_NORTH_SEA from the drop down menu in the INPUT FEATURES section. Once selected it will appear in the FEATURES section immediately below the INPUT FEATURES window. Repeat this for the LAND_NORTH_ SEA_PROJECT_2 data layer. Type C:\GIS_ EXERCISES\POLYGON_GRID_NORTH_SEA_NO_ LAND in the OUTPUT FEATURE CLASS section. In the JOIN ATTRIBUTES section, select ALL from the drop down menu. Next, click OK.

You now want to remove any features from the new data layer which fall on land. To do this, go to SELECTION on the main menu bar and select SELECT BY LOCATION. In the SELECT BY LOCATION window, make sure that under SELECTION METHOD select SELECT FEATURES FROM. Under TARGET LAYERS make sure there is a tick beside POLYGON_GRID_NORTH_ SEA_NO_LAND and no other data layers. Under SOURCE LAYER select LAND_NORTH_ SEA_PROJECT_2. Under SPATIAL SELECTION METHOD FOR TARGET FEATURE(S) select ARE WITHIN THE SOURCE LAYER FEATURE. Now click OK.

2. Remove any new polygons which fall on land

On the EDITOR tool bar, click the EDITOR button and select START EDITING. An EDITOR window should now appear on the right hand side of the MAP window. If the EDITOR window does not appear automatically, you can open it by clicking on the EDITOR button and selecting EDITING WINDOWS> CREATE FEATURES. In this window, select POLYGON_GRID_NORTH_SEA_NO_LAND by clicking on its name once. Next, click on EDIT on the main menu bar and select DELETE. Click on the EDITOR button on the EDITOR tool bar again and select STOP EDITING. When the SAVE window opens click on YES to save the edits.

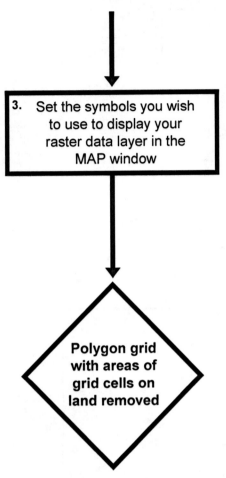

3. Set the symbols you wish to use to display your raster data layer in the MAP window

Right click on the name POLYGON_GRID_ NORTH_SEA_NO_LAND in the TABLE OF CONTENTS window and select PROPERTIES. Next, click on the SYMBOLOGY tab of the LAYER PROPERTIES window. In the left hand portion of the window, select FEATURES> SINGLE SYMBOL. Next, click on the button with the coloured rectangle on it to open the SYMBOL SELECTOR window. Select HOLLOW in the left hand section of the window. Click the OK button to close the SYMBOL SELECTOR window. Finally, click OK to close the LAYER PROPERTIES window.

Polygon grid with areas of grid cells on land removed

At the end of this step, remove the data layer POLYGON_GRID_NORTH_SEA by right-clicking on its name in the TABLE OF CONTENTS window and selecting REMOVE. Now click on the name POLYGON_GRID_NORTH_SEA_NO_LAND in the TABLE OF CONTENTS window and while holding down the left mouse button, drag it up to the top of the list of data layers. Your TABLE OF CONTENTS window should now look like the image on the top of the next page.

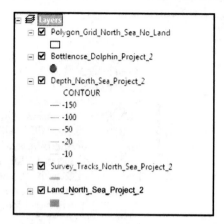

While the contents of your MAP window should look like this:

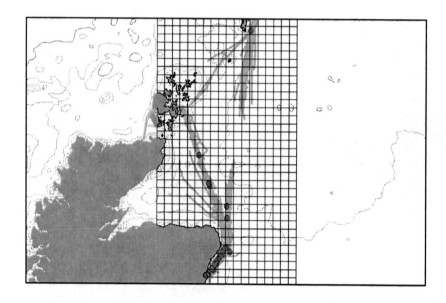

STEP 7: CALCULATE THE LENGTH OF SURVEY TRACKS IN EACH POLYGON GRID CELL:

The next step in this exercise is to divide the survey tracks into sections which fall in each grid cell and then calculate the lengths of all these sections. This is done using the INTERSECT tool as outlined below. Once this has been calculated this information can be joined to the attribute table of the polygon grid data layer using a SPATIAL JOIN.

The instructions for this step are based on instruction sets called *How to split lines or polygons in a data layer into two or more parts*, *How to add information on the length of lines or the area of polygons to the attribute table of a data layer*, *How to remove features from an existing data layer*, *How to join information in the attribute tables of different data layers together based on their spatial relationships (spatial join)* and *How to change the display symbols for a data layer*. The generic versions of these instruction sets can be found in *An Introduction To Using GIS In Marine Biology*.

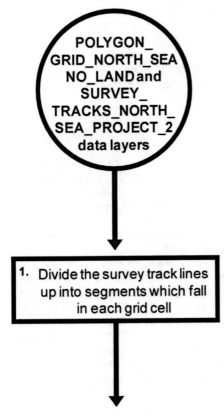

In the TOOLBOX window, select ANALYSIS TOOLS> OVERLAY> INTERSECT. This will open the INTERSECT tool window. In the INTERSECT tool window, select SURVEY_TRACKS_NORTH_SEA_ PROJECT_2 from the drop down menu in the INPUT FEATURES section of the window. Once selected, it will appear in the FEATURES section immediately below the INPUT FEATURES section. Repeat this for POLYGON_GRID_NORTH_SEA_NO_LAND. Type C:\GIS_EXERCISES\SURVEY_TRACKS_INTERSECT in the OUTPUT FEATURE CLASS section of the window. In the JOIN ATTRIBUTES section, select ALL from the drop down menu. In the OUTPUT TYPE section, select INPUT from the drop down menu. Finally click OK to close the INTERSECT tool window.

113

```
┌─────────────────────────────┐
│ 2.  Add field called LENGTH │
│     and calculate length of each │
│     newly created segment   │
└─────────────────────────────┘
```

Right click on the name SURVEY_ TRACKS_INTERSECT in the TABLE OF CONTENTS window, and select OPEN ATTRIBUTE TABLE. Click on the OPTIONS button at the top left corner of the TABLE window and select ADD FIELD. Name the field 'Length', and select SHORT INTEGER for the type. Type in 16 for PRECISION. Now click OK. Next, right-click field name LENGTH in the ATTRIBUTE TABLE window and select CALCULATE GEOMETRY. If a window appears warning you that you are editing a data layer outside of an edit session, click YES, and carry on. If this window does not appear, this is OK. In the CALCULATE GEOMETRY window, for PROPERTY select LENGTH. For COORDINATE SYSTEM select USE COORDINATE SYSTEM OF THE DATA SOURCE. In UNITS select METRES. Click OK.

You now want to remove any line features from the data layer which have a zero length. To do this, go to SELECTION on the main menu bar and select SELECT BY ATTRIBUTES. In the SELECT BY ATTRIBUTES window, select SURVEY_TRACKS_ INTERSECT for LAYER and CREATE A NEW SELECTION for METHOD. Double click on the field name LENGTH to add it to the lower window. Now click on the equals (=) sign to add it to the lower window before typing in a space followed by the number zero (0). This will result in the expression "LENGTH" = 0 appearing in the lower window. Now click OK.

```
┌─────────────────────────────┐
│ 3.  Remove any new lines    │
│     which have a zero length │
└─────────────────────────────┘
```

On the EDITOR tool bar, click the EDITOR button and select START EDITING. The EDITOR window should now appear at the right hand side of the MAP window. If the EDITOR window does not appear automatically, you can open it by clicking on the EDITOR button and selecting EDITING WINDOWS> CREATE FEATURES. If the name of the data layer SURVEY_TRACKS_INTERSECT appears in this window, select it by clicking on it once. If it doesn't, click on the ORGANIZE TEMPLATES button at the top of the EDITOR window (it should be second from the left). Click on the NEW TEMPLATE button at the top of the ORGANIZE FEATURE TEMPLATE window and select the data layer SURVEY_TRACKS_ INTERSECT, then click on the FINISH button. This data layer will now appear in the EDITOR window and you will be able to select it. Next, click on EDIT on the main menu bar and select DELETE. Click on the EDITOR button on the EDITOR tool bar again and select STOP EDITING. When the SAVE window opens click on YES to save the edits.

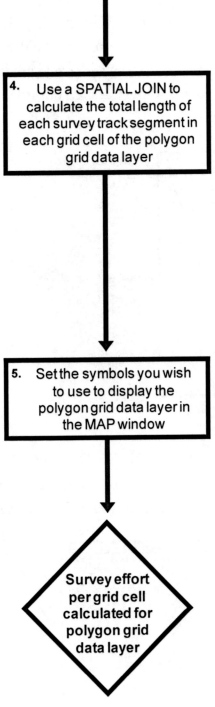

4. Use a SPATIAL JOIN to calculate the total length of each survey track segment in each grid cell of the polygon grid data layer

5. Set the symbols you wish to use to display the polygon grid data layer in the MAP window

Survey effort per grid cell calculated for polygon grid data layer

In the TOOLBOX window, select ANALYSIS TOOLS> OVERLAY> SPATIAL JOIN. In the SPATIAL JOIN window, select the POLYGON_GRID_NORTH_ SEA_NO_LAND in the TARGET FEATURES section of the SPATIAL JOIN window. Select SURVEY_ TRACKS_INTERSECT in the JOIN FEATURES section. Type C:\GIS_EXERCISES\POLYGON_ GRID_NORTH_SEA_SURVEY_EFFORT in the OUTPUT FEATURE CLASS section.

In the JOIN OPERATION (OPTIONAL) section of the window select JOIN_ONE_TO_ONE. In the FIELD MAP OF JOIN FEATURES (OPTIONAL) section, right click on the field called LENGTH and select MERGE RULE> SUM. Remove all fields except FID_POLYGO (LONG), LENGTH (DOUBLE) and CELL_ID_NO (LONG). This is done by right-clicking on a field's name and selecting DELETE. Scroll down until you see the section called MATCH OPTION (OPTIONAL) and select CONTAINS from its drop down menu. Finally, click OK.

Right click on the name of your newly created polygon grid data layer (POLYGON_GRID_NORTH_SEA_ SURVEY_EFFORT in the TABLE OF CONTENTS window and select PROPERTIES. Next, click on the SYMBOLOGY tab of the LAYER PROPERTIES window. In the left hand portion of the LAYER PROPERTIES window, select QUANTITIES> GRADUATED COLOURS. From the drop down menu beside where is says COLOR RAMP select the option that goes from white to black (you may have to scroll down to find it). Next to VALUE select LENGTH. Under RANGE, click on the top line and type in 0, click on the next line and type in 10000. Type in 50000 for the next, 100000 for the next and 5000000 for the last one. Finally, click OK to close the LAYER PROPERTIES window.

Once you have finished working through this flow diagram, remove the data layers SURVEY_TRACKS_INTERSECT and POLYGON_GRID_NORTH_SEA_NO_ LAND from your GIS project by right-clicking on their names in the TABLE OF CONTENTS window and selecting REMOVE. Finally, change the order of the data layers by clicking on their names in the TABLE OF CONTENTS window and, while holding the mouse button down, dragging them upwards or downwards to their desired position so that your TABLE OF CONTENTS window looks like this:

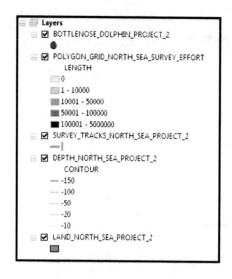

The ATTRIBUTE TABLE for the data layer POLYGON_GRID_NORTH_SEA_ SURVEY_EFFORT should now either look like the figure below or, more likely, the one at the top of the next page. (**NOTE:** The OBJECT ID field may be called the FID field.)

OBJECTID	Shape	Join_Count	TARGET_FID	FID_polygo	CELL_ID_NO	LENGTH	Shape_Length	Shape_Area
1	Polygon	0	0	0	1	0	40000	100000000
2	Polygon	0	1	1	2	0	40000	100000000
3	Polygon	0	2	2	3	0	40000	100000000
4	Polygon	0	3	3	4	0	40000	100000000
5	Polygon	0	4	4	5	0	40000	100000000
6	Polygon	0	5	5	6	0	40000	100000000
7	Polygon	0	6	6	7	0	40000	100000000
8	Polygon	0	7	7	8	0	40000	100000000
9	Polygon	0	8	8	9	0	40000	100000000
10	Polygon	0	9	9	10	0	40000	100000000
11	Polygon	0	10	10	11	0	47692.904314	98230232.244097
12	Polygon	0	11	11	12	0	89331.215494	37940833.23534
13	Polygon	446	12	12	13	1727287	44441.675802	98210513.217431
14	Polygon	0	13	13	14	0	40000	100000000
15	Polygon	0	14	14	15	0	40000	100000000
16	Polygon	0	15	15	16	0	40000	100000000
17	Polygon	0	16	16	17	0	40000	100000000
18	Polygon	0	17	17	18	0	40000	100000000
19	Polygon	0	18	18	19	0	40000	100000000
20	Polygon	0	19	19	20	0	40000	100000000

OBJECTID *	Shape *	Join_Count	TARGET_FID	FID_polygo	CELL_ID_NO	LENGTH
1	Polygon	0	0	0	1	0
2	Polygon	0	1	1	2	0
3	Polygon	0	2	2	3	0
4	Polygon	0	3	3	4	0
5	Polygon	0	4	4	5	0
6	Polygon	0	5	5	6	0
7	Polygon	0	6	6	7	0
8	Polygon	0	7	7	8	0
9	Polygon	0	8	8	9	0
10	Polygon	0	9	9	10	0
11	Polygon	0	10	10	11	0
12	Polygon	0	11	11	12	0
13	Polygon	446	12	12	13	1727287
14	Polygon	0	13	13	14	0
15	Polygon	0	14	14	15	0
16	Polygon	0	15	15	16	0
17	Polygon	0	16	16	17	0
18	Polygon	0	17	17	18	0
19	Polygon	0	18	18	19	0
20	Polygon	0	19	19	20	0

Finally, click on the box to the left of the data layer called SURVEY_TRACKS_NORTH_SEA_PROJECT_2 in the TABLE OF CONTENTS window so that the tick in it disappears and this data layer is no longer displayed.

The contents of your MAP window should look like this:

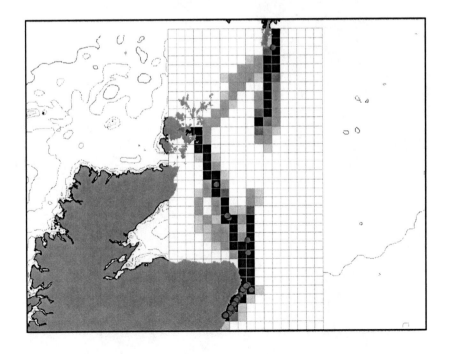

STEP 8: CALCULATE THE NUMBER OF BOTTLENOSE DOLPHINS RECORDED IN EACH GRID CELL:

The final data which must be added to the attribute table of the polygon grid data layer is information on the number of bottlenose dolphins recorded in each grid cell. This is also added using a SPATIAL JOIN to calculate the total number of dolphins for all sightings which fall within each grid cell. The flow diagram for this step can be found on the next page.

The instructions for this step are based on instruction sets called *How to join information in the attribute tables of different data layers together based on their spatial relationships (spatial join)* and *How to change the display symbols for a data layer* from *An Introduction To Using GIS In Marine Biology*.

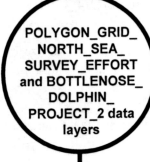

POLYGON_GRID_NORTH_SEA_SURVEY_EFFORT and BOTTLENOSE_DOLPHIN_PROJECT_2 data layers

1. Use a SPATIAL JOIN to calculate the total number of bottlenose dolphin recorded in each grid cell of the polygon grid data layer

In the TOOLBOX window, select ANALYSIS TOOLS> OVERLAY> SPATIAL JOIN. In the SPATIAL JOIN window, select the POLYGON_GRID_NORTH_SEA_SURVEY_EFFORT in the TARGET FEATURES section. Select BOTTLENOSE_DOLPHIN_PROJECT_2 in the JOIN FEATURES section. Type C:\GIS_EXERCISES\POLYGON_GRID_NORTH_SEA_WITH_SIGHTINGS in the OUTPUT FEATURE CLASS section.

In the JOIN OPERATION (OPTIONAL) section of the window select JOIN_ONE_TO_ONE. In the FIELD MAP OF JOIN FEATURES (OPTIONAL) section, right click on the field called NUMBER (DOUBLE) and select MERGE RULE> SUM. Remove all fields except FID_POLYGO (LONG), LENGTH (DOUBLE), CELL_ID_NO (LONG) and NUMBER (DOUBLE). This is done by right-clicking on a field's name and selecting DELETE. Scroll down until you see the section called MATCH OPTION (OPTIONAL) and select CONTAINS from its drop down menu. Finally, click OK.

Polygon grid data layer with a field in the attribute table with the total number of bottlenose dolphins recorded in each grid cell

119

At the end of this step, remove the data layer called POLYGON_GRID_NORTH_SEA _SURVEY_EFFORT by right-clicking on its name in the TABLE OF CONTENTS window and selecting REMOVE.

Finally, open the TABLE window for the data layer POLYGON_ GRID_NORTH_SEA_WITH_SIGHTINGS by right-clicking on its name in the TABLE OF CONTENTS window and selecting OPEN ATTRIBUTE TABLE. The attribute table should look like this:

FID	Shape *	Join_Count	TARGET_FID	FID_polygo	CELL_ID_NO	LENGTH	Number
0	Polygon	0	1	0	1	0	0
1	Polygon	0	2	1	2	0	0
2	Polygon	0	3	2	3	0	0
3	Polygon	0	4	3	4	0	0
4	Polygon	0	5	4	5	0	0
5	Polygon	0	6	5	6	0	0
6	Polygon	0	7	6	7	0	0
7	Polygon	0	8	7	8	0	0
8	Polygon	0	9	8	9	0	0
9	Polygon	0	10	9	10	0	0
10	Polygon	0	11	10	11	0	0
11	Polygon	0	12	11	12	0	0
12	Polygon	0	13	12	13	1727287	0
13	Polygon	0	14	13	14	0	0
14	Polygon	0	15	14	15	0	0
15	Polygon	0	16	15	16	0	0
16	Polygon	0	17	16	17	0	0
17	Polygon	0	18	17	18	0	0
18	Polygon	0	19	18	19	0	0
19	Polygon	0	20	19	20	0	0
20	Polygon	0	21	20	21	0	0

Now close the attribute table.

STEP 9: CALCULATE THE ABUNDANCE PER UNIT EFFORT IN EACH GRID CELL:

You now have all the information in the attribute table of the polygon grid data layer which you would need to calculate the abundance of bottlenose dolphins per kilometre of survey effort in each grid cell. This will be calculated by adding a new field and using the FIELD CALCULATOR tool to divide the number of bottlenose dolphins recorded in each grid cell by the amount of survey effort.

These instructions are based on instruction sets called *How to add a new field to an attribute table*, *How to use the field calculator tool to fill in values in a new field* and *How to change the display symbols for a data layer* from *An Introduction To Using GIS In Marine Biology*.

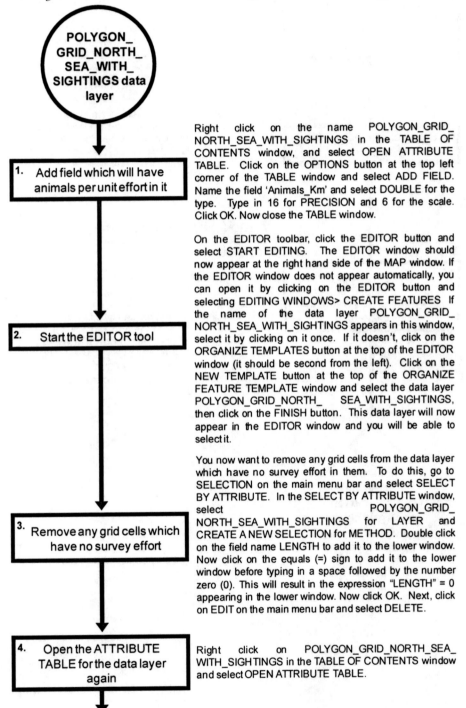

POLYGON_ GRID_NORTH_ SEA_WITH_ SIGHTINGS data layer

1. Add field which will have animals per unit effort in it

Right click on the name POLYGON_GRID_ NORTH_SEA_WITH_SIGHTINGS in the TABLE OF CONTENTS window, and select OPEN ATTRIBUTE TABLE. Click on the OPTIONS button at the top left corner of the TABLE window and select ADD FIELD. Name the field 'Animals_Km' and select DOUBLE for the type. Type in 16 for PRECISION and 6 for the scale. Click OK. Now close the TABLE window.

2. Start the EDITOR tool

On the EDITOR toolbar, click the EDITOR button and select START EDITING. The EDITOR window should now appear at the right hand side of the MAP window. If the EDITOR window does not appear automatically, you can open it by clicking on the EDITOR button and selecting EDITING WINDOWS> CREATE FEATURES If the name of the data layer POLYGON_GRID_ NORTH_SEA_WITH_SIGHTINGS appears in this window, select it by clicking on it once. If it doesn't, click on the ORGANIZE TEMPLATES button at the top of the EDITOR window (it should be second from the left). Click on the NEW TEMPLATE button at the top of the ORGANIZE FEATURE TEMPLATE window and select the data layer POLYGON_GRID_NORTH_ SEA_WITH_SIGHTINGS, then click on the FINISH button. This data layer will now appear in the EDITOR window and you will be able to select it.

3. Remove any grid cells which have no survey effort

You now want to remove any grid cells from the data layer which have no survey effort in them. To do this, go to SELECTION on the main menu bar and select SELECT BY ATTRIBUTE. In the SELECT BY ATTRIBUTE window, select POLYGON_GRID_ NORTH_SEA_WITH_SIGHTINGS for LAYER and CREATE A NEW SELECTION for METHOD. Double click on the field name LENGTH to add it to the lower window. Now click on the equals (=) sign to add it to the lower window before typing in a space followed by the number zero (0). This will result in the expression "LENGTH" = 0 appearing in the lower window. Now click OK. Next, click on EDIT on the main menu bar and select DELETE.

4. Open the ATTRIBUTE TABLE for the data layer again

Right click on POLYGON_GRID_NORTH_SEA_ WITH_SIGHTINGS in the TABLE OF CONTENTS window and select OPEN ATTRIBUTE TABLE.

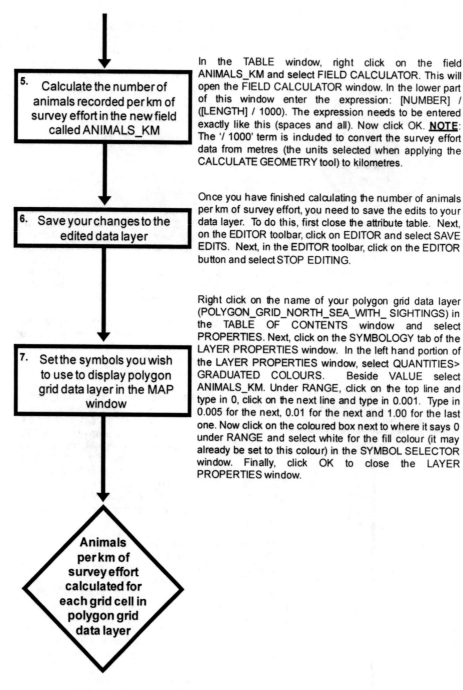

5. Calculate the number of animals recorded per km of survey effort in the new field called ANIMALS_KM

In the TABLE window, right click on the field ANIMALS_KM and select FIELD CALCULATOR. This will open the FIELD CALCULATOR window. In the lower part of this window enter the expression: [NUMBER] / ([LENGTH] / 1000). The expression needs to be entered exactly like this (spaces and all). Now click OK. **NOTE**: The '/ 1000' term is included to convert the survey effort data from metres (the units selected when applying the CALCULATE GEOMETRY tool) to kilometres.

6. Save your changes to the edited data layer

Once you have finished calculating the number of animals per km of survey effort, you need to save the edits to your data layer. To do this, first close the attribute table. Next, on the EDITOR toolbar, click on EDITOR and select SAVE EDITS. Next, in the EDITOR toolbar, click on the EDITOR button and select STOP EDITING.

7. Set the symbols you wish to use to display polygon grid data layer in the MAP window

Right click on the name of your polygon grid data layer (POLYGON_GRID_NORTH_SEA_WITH_ SIGHTINGS) in the TABLE OF CONTENTS window and select PROPERTIES. Next, click on the SYMBOLOGY tab of the LAYER PROPERTIES window. In the left hand portion of the LAYER PROPERTIES window, select QUANTITIES> GRADUATED COLOURS. Beside VALUE select ANIMALS_KM. Under RANGE, click on the top line and type in 0, click on the next line and type in 0.001. Type in 0.005 for the next, 0.01 for the next and 1.00 for the last one. Now click on the coloured box next to where it says 0 under RANGE and select white for the fill colour (it may already be set to this colour) in the SYMBOL SELECTOR window. Finally, click OK to close the LAYER PROPERTIES window.

Animals per km of survey effort calculated for each grid cell in polygon grid data layer

Once you have calculated the recorded abundance of bottlenose dolphin per kilometre of survey effort per grid cell, you need to turn off all the other data layers with the exception of LAND_NORTH_SEA, POLYGON_GRID_NORTH_SEA_WITH SIGHTINGS and DEPTH_NORTH_SEA.

The ATTRIBUTE TABLE for POLYGON_GRID_NORTH_SEA_WITH_ SIGHTINGS should now look like this:

FID	Shape *	Join_Count	TARGET_FID	FID_polygo	CELL_ID_NO	LENGTH	Number	Animals_km
0	Polygon	0	13	12	13	1727287	0	0
1	Polygon	0	30	29	30	2852	0	0
2	Polygon	0	31	30	31	1748827	0	0
3	Polygon	0	48	47	48	96869	0	0
4	Polygon	1	49	48	49	1626920	25	0.015366
5	Polygon	0	64	63	64	1096	0	0
6	Polygon	0	65	64	65	32774	0	0
7	Polygon	0	66	65	66	814236	0	0
8	Polygon	0	67	66	67	1038366	0	0
9	Polygon	0	81	80	81	12738	0	0
10	Polygon	0	82	81	82	25747	0	0
11	Polygon	0	83	82	83	15115	0	0
12	Polygon	0	84	83	84	1318625	0	0
13	Polygon	0	85	84	85	184766	0	0
14	Polygon	0	97	96	97	5021	0	0
15	Polygon	0	98	97	98	11561	0	0
16	Polygon	0	99	98	99	23209	0	0
17	Polygon	0	100	99	100	12789	0	0
18	Polygon	0	101	100	101	2457	0	0
19	Polygon	0	102	101	102	1333778	0	0
20	Polygon	0	103	102	103	58460	0	0

Your TABLE OF CONTENTS window should now look like this:

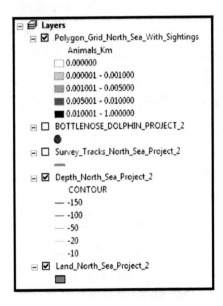

123

And the contents of your MAP window should look like this:

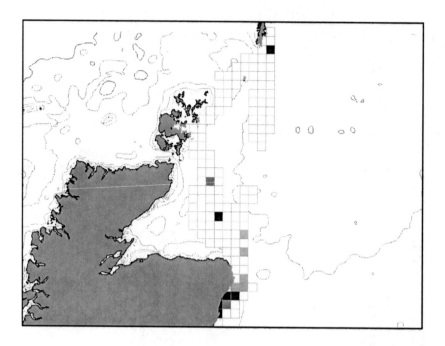

Optional extra:

If you wish to have more practice at calculating the abundance of animals per unit of survey effort in grid cells for a specific study area, you can repeat this exercise for harbour porpoises. This would involve exactly the same steps, with the exception that it would be harbour porpoise sightings locations that would be extracted in step four and used throughout the rest of the exercise. This means that you can either choose to repeat this exercise from the beginning, or you can simply repeat steps four, eight and nine as you already have a polygon grid data layer which contains a measure of survey effort per grid cell in it. If you are going to repeat this exercise from the start, you will need to use slightly different names for the data layers you create when repeating this for harbour porpoises. This is because you will need some of the data layers created in the bottlenose dolphin version of this exercise in exercise five (see below).

Once you have calculated the recorded abundance of harbour porpoises per kilometre of survey effort per grid cell, and you have turned off all the other data layers with the exception of LAND_NORTH_SEA, POLYGON_GRID_NORTH_SEA_WITH SIGHTINGS and DEPTH_NORTH_SEA, the contents of your MAP window should look like this:

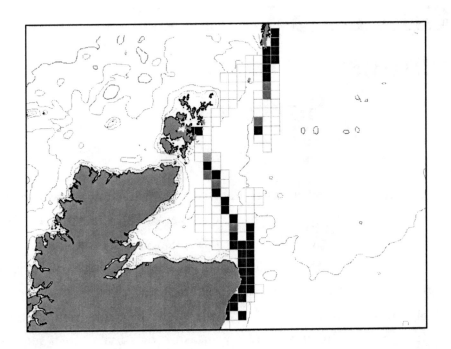

Chapter Seven

Exercise Five: Creating A Polygon Grid Data Layer Of Species Richness Per Unit Effort From Survey Data

In exercise three, you produced a raster data layer which showed how many different species were recorded in each grid cell for the surveys conducted in northeast Scotland. However, while this showed which grid cells had been surveyed, it did not take into account how much survey effort had been conducted in each grid. Therefore, it is hard to tell whether one grid cell really has a higher species richness than another or whether any differences are simply a function of different levels of survey effort between them. In order to investigate this, you need to incorporate a measure of survey effort into your grid in some way. In exercise four, you did something similar with abundance for an individual species using a polygon grid data layer, so you already have a measure of survey effort in each grid cell. Therefore, in this exercise, you will combine the information in the species richness raster data layer you created in exercise three and the polygon grid data layer which has information on survey effort per grid cell created in exercise four to create a polygon grid which contains information on species richness per unit of survey effort. This polygon grid will allow you to tell whether differences in species richness between grid cells is simply a function of effort, or whether they represents real differences in species richness.

For this exercise, you will use two existing data layers. These data layers should all be saved in a folder on your C drive called GIS_EXERCISES prior to starting this exercise, so that it has the address C:\GIS_EXERCISES\.

The data layers needed for this exercise are:

1. Polygon_Grid_North_Sea_With_Sightings.shp: This is the final polygon grid data layer created in exercise four. The attribute table for the data layer contains information on the number of bottlenose dolphins recorded in each polygon grid cell and the total length of survey track sections in it. It is in the custom transverse mercator projection used for other exercises in this book and is based on the WGS 1984 datum.

2. SP_Richness.grd: This is a raster data layer of species richness created in exercise three. Its cell size and its extent are the same as that of Polygon_Grid _North_Sea_With_Sightings.shp, but it is a raster data layer rather than a polygon grid data layer. Again, it is in the custom transverse mercator projection used for other exercises in this book and uses the WGS 1984 datum.

Start the ArcGIS 10.1 software by opening the ArcMap module and open the final version of the GIS project created in exercise four. Next, click on FILE from the main menu bar area, select SAVE As and save it as EXERCISE_FIVE in the folder C:\GIS_EXERCISES\. Next, remove all the existing data layers with the exception of POLYGON_GRID_NORTH_SEA_WITH_SIGHTINGS, LAND_NORTH_SEA_ PROJECT_2 and DEPTH_ NORTH_SEA_ PROJECT_2. The unwanted data layers can be removed by right-clicking on their names in the TABLE OF CONTENTS window and selecting REMOVE.

Thus, at the start of this exercise, your TABLE OF CONTENTS window should look like this:

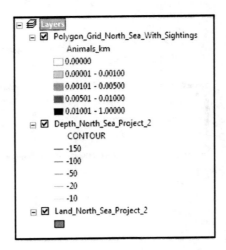

And the contents of your MAP window should look like this:

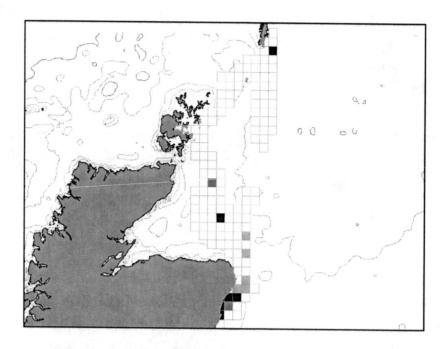

Summary Flow Diagram For Creating A Polygon Grid Data Layer With Abundance Per Unit Effort From Survey Data

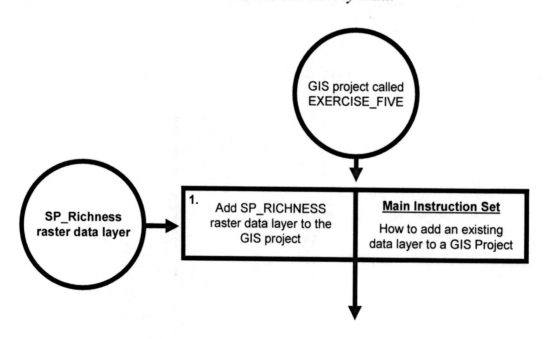

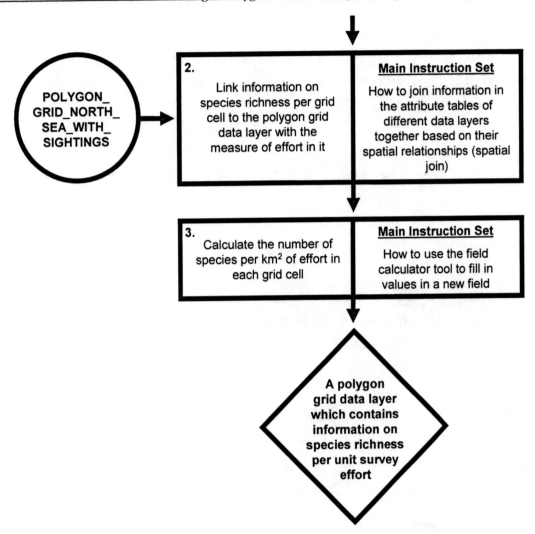

Once you have familiarised yourself with this summary flow diagram, you will need to read through the instructions for the first step before working through them. When you have completed the first step, read through the instructions again and make sure that you have completed it properly. It is important to do this at this stage as you need to use the results of one step as the starting point for the next, so it is important to make sure that you have completed it correctly at this stage. In particular, it is much easier to spot where you have gone wrong at the end of an individual step, rather than trying to work it out later when you get stuck. Once you have successfully completed the first step, move onto the second step and repeat this process, and so on until you have completed all the steps in the summary flow diagram.

At various points, images of the contents of the MAP window, the LAYOUT window, the TABLE OF CONTENTS window and/or the TABLE window will be provided so that you have an idea of what your GIS project should look like at specific points as you progress through this exercise.

Instruction Sets For The Individual Steps Identified In The Summary Flow Diagram:

STEP 1: ADD SP_RICHNESS RASTER DATA LAYER TO THE GIS PROJECT:

In this step, you will add the species richness raster data layer you created in exercise three to the GIS project for this exercise. Once it has been added, you will need to check that it has the right projection/coordinate system assigned to it. This may involve assigning a projection/coordinate system to it.

This instruction set is based on one called *How to add an existing raster data layer to a GIS project* and *How to set the projection/coordinate system for a data layer so that it plots properly over other data layers* and *How to change the display symbols for a data layer*. The generic versions of these instruction sets can be found in *An Introduction To Using GIS In Marine Biology*.

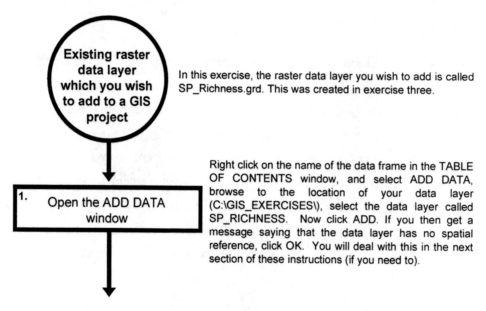

Existing raster data layer which you wish to add to a GIS project

In this exercise, the raster data layer you wish to add is called SP_Richness.grd. This was created in exercise three.

1. Open the ADD DATA window

Right click on the name of the data frame in the TABLE OF CONTENTS window, and select ADD DATA, browse to the location of your data layer (C:\GIS_EXERCISES\), select the data layer called SP_RICHNESS. Now click ADD. If you then get a message saying that the data layer has no spatial reference, click OK. You will deal with this in the next section of these instructions (if you need to).

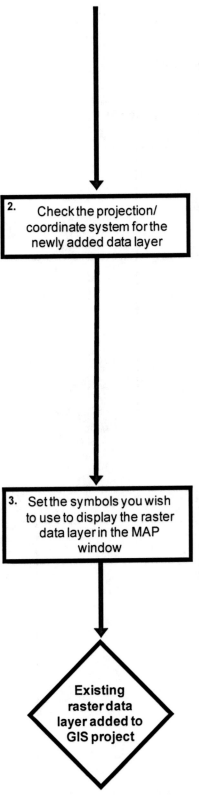

2. Check the projection/ coordinate system for the newly added data layer

3. Set the symbols you wish to use to display the raster data layer in the MAP window

Existing raster data layer added to GIS project

Whenever you add a data layer to a GIS project, you should always check that is has a projection/coordinate system assigned to it, and look at what this projection/coordinate system is. This is so that you know whether you will need to assign a projection/coordinate system to it, or transform it into a different projection/coordinate system before you can use it in your GIS project. To check the projection/coordinate system of your newly added data layer, in the TOOLBOX window, go to DATA MANAGEMENT TOOLS> PROJECTIONS AND TRANSFORMATIONS> DEFINE PROJECTION. Select SP_RICHNESS from the drop down menu in the top window. Check its projection/coordinate system in the lower window. This should be GCS_ WGS_1984. If it says this, click CANCEL, and move onto stage three.

If it says UNKNOWN, click on the button at the right hand end of this section of the window to open the SPATIAL REFERENCE PROPERTIES window. In this window, click on the ADD COORDINATE SYSTEM button (which is in the top right hand corner and has a picture of a globe on it) and select IMPORT. Now browse to and select POLYGON_GRID_NORTH_ SEA_WITH_SIGHTINGS, and then click on the ADD button. This will import the custom transverse mercator projection/coordinate system being used in this exercise (that is called NORTH SEA). Now click OK to close the SPATIAL REFERENCE PROPERTIES window and finally click OK to close the DEFINE PROJECTION window.

Right click on the name of your raster data layer (SP_RICHNESS) in the TABLE OF CONTENTS window and select PROPERTIES. Next, click on the SYMBOLOGY tab of the LAYER PROPERTIES window. In the left hand portion of the LAYER PROPERTIES window, select UNIQUE VALUES. Double click on the coloured rectangle next to 0 and select 10% grey. Repeat for the other values, selecting 30% grey for 1, 50% grey for 2, 70% grey for 3, 80% grey for 4 and black for 5. Finally, click OK to close the LAYER PROPERTIES window.

At the end of this step, you will need to move the SP_RICHNESS raster data layer to the top by clicking on its name in the TABLE OF CONTENTS window and, while holding the left mouse button down, dragging it to its required position.

Your TABLE OF CONTENTS window should now look like this:

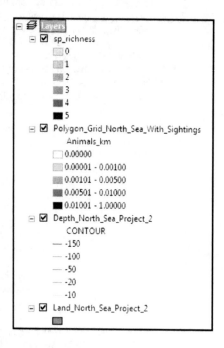

The contents of your MAP window should now look like this:

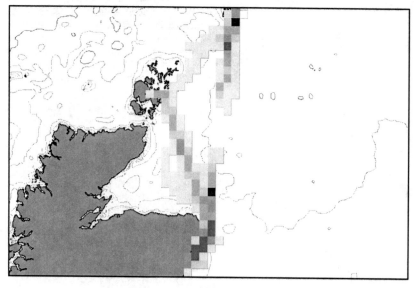

STEP 2: LINK INFORMATION ON SPECIES RICHNESS PER GRID CELL TO THE POLYGON GRID DATA LAYER WITH THE MEASURE OF EFFORT IN IT:

In order to be able to calculate species richness per unit of survey effort, you first need to get the information on species richness into the attribute table of the polygon grid data layer which has the measure of effort in it (in this case POLYGON_GRID_NORTH_ SEA_WITH_SIGHTINGS). The quickest way to do this is to convert the species richness raster data layer into a point data layer which can then be joined to the polygon data layer using a spatial join.

The instruction set for this step is based on ones called *How to join information in the attribute tables of different data layers together based on their spatial relationships (spatial join)* and *How to change the display symbols for a data layer*. Generic versions of these instruction sets can be found in *An Introduction To Using GIS In Marine Biology*. In addition, it includes information from chapter eighteen on converting data layers between data layer types.

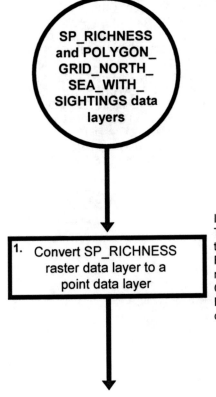

SP_RICHNESS and POLYGON_ GRID_NORTH_ SEA_WITH_ SIGHTINGS data layers

1. Convert SP_RICHNESS raster data layer to a point data layer

In the TOOLBOX window, select CONVERSION TOOLS> FROM RASTER> RASTER TO POINT. In the RASTER TO POINT window, for INPUT RASTER select SP_RICHNESS from the drop down menu. For FIELD (OPTIONAL) select VALUE. In OUTPUT POINT FEATUTES, type in C:\GIS_ EXERCISES\SP_RICHNESS_POINTS. Finally, click OK to create the point data layer.

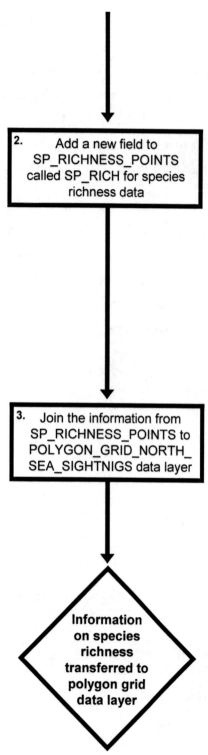

2. Add a new field to SP_RICHNESS_POINTS called SP_RICH for species richness data

3. Join the information from SP_RICHNESS_POINTS to POLYGON_GRID_NORTH_ SEA_SIGHTNIGS data layer

Information on species richness transferred to polygon grid data layer

Right click on the name SP_RICHNESS_POINTS in the TABLE OF CONTENTS window, and select OPEN ATTRIBUTE TABLE. In the TABLE window, click on the OPTIONS in the top left corner and select ADD FIELD. Call the new field SP_RICH and select SHORT INTEGER for the type. Enter 16 for PRECISION. Now click OK.

Right click on the field name SP_RICH and select FIELD CALCULATOR. When a warning appears, click YES. In the FIELD CALCULATOR window, double click on the field GRID_CODE to add it to the lower window, and then click OK. This will transfer the information from the GRID_CODE field to the one called SP_RICH.

Finally, right click on the field name GRID_CODE and select DELETE FIELD. This will delete this field from the attribute table. This is done so that you have your information on the number of species recorded per grid cell in a field with a representative name and not a generic one. Now close the TABLE window.

In the TOOLBOX window, select ANALYSIS TOOLS> OVERLAY> SPATIAL JOIN. In the SPATIAL JOIN window, select POLYGON_GRID_NORTH_SEA_WITH_SIGHTING S as the TARGET FEATURE. For JOIN FEATURES select SP_RICHNESS_POINTS. For OUTPUT FEATURE CLASS type in C:\GIS_ EXERCISES\POLYGON_GRID_NORTH_SEA_WIT H_RICHNESS. For JOIN OPERATION (OPTIONAL) select JOIN_ONE_TO_ONE. In the FIELD MAP OF JOIN FEATURES (OPTIONAL) delete all the fields except FID_POLYGO, CELL_ID_NO (LONG), LENGTH (DOUBLE), ANIMALS_KM (DOUBLE) and SP_RICH (DOUBLE). Fields are deleted by right-clicking on their name and selecting DELETE. Finally, click OK.

Your TABLE OF CONTENTS window should now look like this:

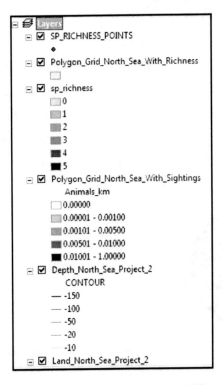

The attribute table for POLYGON_GRID_NORTH_SEA_WITH_RICHNESS should look like the table below. If it does not look like this (and especially if the join count values are zero), remove the SP_RICHNESS_POINTS data layer from your GIS project. Next add it back in and repeat stage three of this step.

FID	Shape *	Join_Count	TARGET_FID	FID_polygo	CELL_ID_NO	LENGTH	Animals_km	SP_RICH
0	Polygon	1	0	12	13	1727287	0	3
1	Polygon	1	1	29	30	2852	0	0
2	Polygon	1	2	30	31	1748627	0	3
3	Polygon	1	3	47	48	96869	0	1
4	Polygon	1	4	48	49	1626920	0.015366	5
5	Polygon	1	5	63	64	1096	0	0
6	Polygon	1	6	64	65	32774	0	0
7	Polygon	1	7	65	66	614236	0	2
8	Polygon	1	8	66	67	1038366	0	2
9	Polygon	1	9	80	81	12738	0	0
10	Polygon	1	10	81	82	25747	0	0
11	Polygon	1	11	82	83	15115	0	1
12	Polygon	1	12	83	84	1318525	0	3
13	Polygon	1	13	84	85	184766	0	1
14	Polygon	1	14	96	97	5021	0	0
15	Polygon	1	15	97	98	11561	0	0
16	Polygon	1	16	98	99	23209	0	0
17	Polygon	1	17	99	100	12789	0	0
18	Polygon	1	18	100	101	2457	0	0
19	Polygon	1	19	101	102	1333776	0	4
20	Polygon	1	20	102	103	58480	0	0

STEP 3: CALCULATE THE NUMBER OF SPECIES PER KM OF EFFORT IN EACH GRID CELL:

The final step in this exercise is to calculate the number of species per kilometre (km) of survey effort in each grid cell. This is done by creating a new field in the attribute table of the polygon grid data layer and using the FIELD CALCULATOR tool to calculate this value from the information in the SP_RICH field and the LENGTH field. This is done as outlined below. This instruction set is based on ones called *How to use the field calculator tool to transfer data between fields in an attribute table* and *How to change the display symbols for a data layer* in *An Introduction To Using GIS In Marine Biology*.

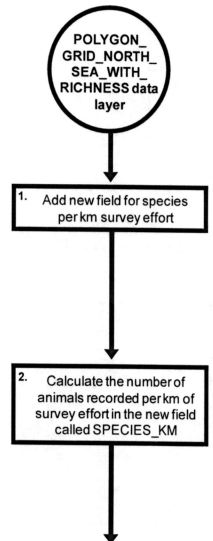

Right click on the name POLYGON_GRID_ NORTH_SEA_WITH_RICHNESS in the TABLE OF CONTENTS window, and select OPEN ATTRIBUTE TABLE. Click on the OPTIONS button at the top left corner of the TABLE window and select ADD FIELD. Name the field 'Species_km' and select DOUBLE for the type. Type in 16 for PRECISION and 6 for the scale. Click OK.

Move the TABLE window so that you can see the main menu bar and optional toolbar areas of the main MAP window. Click on the CUSTOMIZE menu in the main menu bar and select TOOLBARS. Make sure there is a tick next to the EDITOR toolbar. In the EDITOR toolbar, click on the EDITOR button and select START EDITING. If a warning window appears, this is okay. Just Click CONTINUE and carry on. A new EDITOR window may then appear (usually on the right hand side of the main MAP window – do not worry if this window does not appear). Now click on the TABLE window and then right click on the field SPECIES_KM and select FIELD CALCUALTOR. This will open the FIELD CALCULATOR window. In the lower part of this window enter the expression: [SP_RICH] / ([LENGTH] / 1000) The expression needs to be entered exactly like this (spaces and all). Finally, click OK to do the calculation.

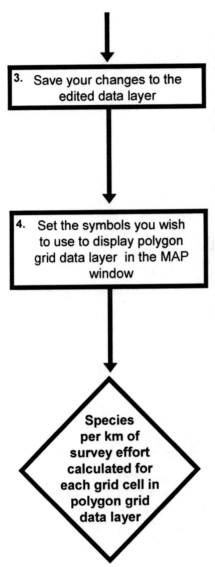

3. Save your changes to the edited data layer

4. Set the symbols you wish to use to display polygon grid data layer in the MAP window

Species per km of survey effort calculated for each grid cell in polygon grid data layer

Once you have finished calculating the number of species per km of survey effort, you need to save the edits to your data layer. To do this, first close the attribute table. Next, on the EDITOR toolbar, click on EDITOR and select SAVE EDITS. Next, in the EDITOR toolbar, click on the EDITOR button and select STOP EDITING.

Right click on the name of your polygon grid data layer (POLYGON_GRID_NORTH_SEA_WITH_ RICHNESS) in the TABLE OF CONTENTS window and select PROPERTIES. Next, click on the SYMBOLOGY tab of the LAYER PROPERTIES window. In the left hand portion of the LAYER PROPERTIES window, select QUANTITIES> GRADUATED COLOURS. For VALUE select SPECIES_KM. For COLOUR RAMP, select a colour ramp going from white to black. Next, click on the numbers under RANGE for each of the five categories in turn and enter the values 0, 0.001, 0.005, 0.01 and 1. Finally, click OK to close the LAYER PROPERTIES window.

Once you have finished this final step, remove the data layers SP_RICHNESS and SP_RICHNESS_POINTS from your GIS project by right-clicking on their names in the TABLE OF CONTENTS window and selecting REMOVE.

The contents of your MAP window should then look like this:

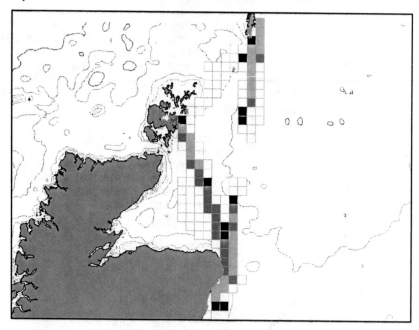

While your TABLE OF CONTENTS window should look like this:

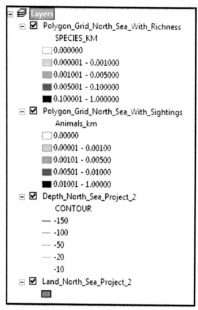

The ATTRIBUTE TABLE for POLYGON_GRID_NORTH_SEA_WITH_ RICHNESS should look like this:

FID	Shape *	Join_Count	TARGET_FID	FID_polygo	CELL_ID_NO	LENGTH	Animals_km	SP_RICH	Species_km
0	Polygon	1	0	12	13	1727287	0	3	0.001737
1	Polygon	1	1	29	30	2852	0	0	0
2	Polygon	1	2	30	31	1748827	0	3	0.001715
3	Polygon	1	3	47	48	96869	0	1	0.010323
4	Polygon	1	4	48	49	1626920	0.015366	5	0.003073
5	Polygon	1	5	63	64	1096	0	0	0
6	Polygon	1	6	64	65	32774	0	0	0
7	Polygon	1	7	65	66	614236	0	2	0.003258
8	Polygon	1	8	66	67	1038366	0	2	0.001925
9	Polygon	1	9	80	81	12738	0	0	0
10	Polygon	1	10	81	82	25747	0	0	0
11	Polygon	1	11	82	83	15115	0	1	0.066159
12	Polygon	1	12	83	84	1318525	0	3	0.002275
13	Polygon	1	13	84	85	184766	0	1	0.005412
14	Polygon	1	14	96	97	5021	0	0	0
15	Polygon	1	15	97	98	11561	0	0	0
16	Polygon	1	16	98	99	23209	0	0	0
17	Polygon	1	17	99	100	12789	0	0	0
18	Polygon	1	18	100	101	2457	0	0	0
19	Polygon	1	19	101	102	1333776	0	4	0.002999
20	Polygon	1	20	102	103	58460	0	0	0

Optional extra:

If you wish to get more experience, you can repeat this process for the SP_ABUNDANCE raster data layer created in the optional extra section of exercise three to create a polygon grid data layer which contains a measure of the total abundance of cetaceans per unit survey effort.

If you use the same display settings used for POLYGON_GRID_NORTH_SEA_ WITH_RICHNESS, when finished the contents of your MAP window should look like the figure at the top of the next page.

139

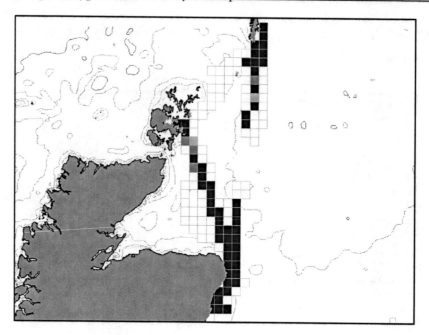